INVESTIGATIONS IN NUMBER, DATA,

Graphs

Changes Over Time

Grade 4

Also appropriate for Grades 5 and 6

Cornelia Tierney
Ricardo Nemirovsky
Amy Shulman Weinberg

Developed at TERC, Cambridge, Massachusetts

Dale Seymour Publications

The *Investigations* curriculum was developed at TERC (formerly Technical Education Research Centers) in collaboration with Kent State University and the State University of New York at Buffalo. The work was supported in part by National Science Foundation Grant No. MDR-9050210. TERC is a nonprofit company working to improve mathematics and science education. TERC is located at 2067 Massachusetts Avenue, Cambridge, MA 02140.

This project was supported, in part, by the
National Science Foundation
Opinions expressed are those of the authors and not necessarily those of the Foundation

This book is published by Dale Seymour Publications, an imprint of the Alternative Publishing Group of Addison-Wesley Publishing Company.

Project Editor: Priscilla Cox Samii
Series Editor: Beverly Cory
Manuscript Editor: Nancy Tune
ESL Consultant: Nancy Sokol Green
Production/Manufacturing Director: Janet Yearian
Production/Manufacturing Coordinator: Barbara Atmore
Design Manager: Jeff Kelly
Design: Don Taka
Illustrations: DJ Simison, Carl Yoshihara
Cover: Bay Graphics
Composition: Publishing Support Services

Copyright © 1995 by Dale Seymour Publications. All rights reserved.
Printed in the United States of America.

 Printed on Recycled Paper

Limited reproduction permission: The publisher grants permission to individual teachers who have purchased this book to reproduce the blackline masters as needed for use with their own students. Reproduction for an entire school or school district or for commercial use is prohibited.

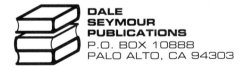

DALE SEYMOUR PUBLICATIONS
P.O. BOX 10888
PALO ALTO, CA 94303

Order number DS21249
ISBN 0-86651-811-8
2 3 4 5 6 7 8 9 10-ML-98 97 96 95 94

TERC

INVESTIGATIONS IN NUMBER, DATA, AND SPACE

Principal Investigator   Susan Jo Russell
Co-Principal Investigator   Cornelia C. Tierney
Director of Research and Evaluation   Jan Mokros

Curriculum Development

Joan Akers
Michael T. Battista
Mary Berle-Carman
Douglas H. Clements
Karen Economopoulos
Ricardo Nemirovsky
Andee Rubin
Susan Jo Russell
Cornelia C. Tierney
Amy Shulman Weinberg

Evaluation and Assessment

Mary Berle-Carman
Abouali Farmanfarmaian
Jan Mokros
Mark Ogonowski
Amy Shulman Weinberg
Tracey Wright
Lisa Yaffee

Teacher Development and Support

Rebecca B. Corwin
Karen Economopoulos
Tracey Wright
Lisa Yaffee

Technology Development

Michael T. Battista
Douglas H. Clements
Julie Sarama Meredith
Andee Rubin

Video Production

David A. Smith

Administration and Production

Amy Catlin
Amy Taber

Cooperating Classrooms for This Unit

Kathleen O'Connell
Lynda Verity
Arlington Public Schools, Arlington, MA

Michele deSilva
Angela Philactos
Jim Sampson
Boston Public Schools, Boston, MA

Susan Wheelright
Fayerweather School, Cambridge, MA

Consultants and Advisors

Elizabeth Badger
Deborah Lowenberg Ball
Marilyn Burns
Ann Grady
Joanne M. Gurry
James J. Kaput
Steven Leinwand
Mary M. Lindquist
David S. Moore
John Olive
Leslie P. Steffe
Peter Sullivan
Grayson Wheatley
Virginia Woolley
Anne Zarinnia

Graduate Assistants

Kent State University:
Joanne Caniglia, Pam DeLong, Carol King

State University of New York at Buffalo:
Rosa Gonzalez, Sue McMillen,
Julie Sarama Meredith, Sudha Swaminathan

CONTENTS

About the *Investigations* Curriculum   1
How to Use This Book   2
About Assessment   7

Changes Over Time
Overview   8
Materials List   10
About the Mathematics in This Unit   11
Preview for the Linguistically Diverse Classroom   12

Unit Preparation: Growing Plants to Graph   14
Preparation Session 1: Sprouting the Seeds   16
Preparation Session 2: Planting the Seeds   18
Preparation Session 3: Measuring the Plants   20

Investigation 1: Graphing Population Changes   22
Sessions 1 and 2: Graphing Home Population Over Time   24
Sessions 3 and 4: Making Group Graphs   30
Sessions 5 and 6 (Excursion): Ins and Outs Number Problems   35

Investigation 2: Ways to Show Change Over Time   42
Sessions 1 and 2: Showing Change Over Time   44

Investigation 3: Telling Stories from Line Graphs   48
Session 1: Making Graph Sketches   50
Session 2: Graphing and Predicting Plant Growth   58
Session 3: Using Line Graphs to Compare Growth   65
Session 4: Graphs, Stories, and Number Sequences   68
Session 5: Interpreting Graphs   71
Sessions 6 and 7: Mystery Graphs   74

Appendix: Ten-Minute Math   82
Appendix: Vocabulary Support for Second-Language Learners   86
Blackline Masters   89
    Family Letter
    Student Sheets 1–9
    Teaching Resources

Teacher Notes

Varieties of Students' Graphs	29
Problems to Solve When Combining Data	34
Strategies for Ins and Outs Problems	41
Graph Sketches of Speed	56
Speed Versus Change in Speed	57
Continuous Versus Discrete Changes	63
Students' Line Graphs	64
Assessment: Telling Stories from Graphs	72

ABOUT THE *INVESTIGATIONS* CURRICULUM

Investigations in Number, Data, and Space is a K–5 mathematics curriculum with four major goals:

- to offer students meaningful mathematical problems
- to emphasize depth in mathematical thinking rather than superficial exposure to a series of fragmented topics
- to communicate mathematics content and pedagogy to teachers
- to substantially expand the pool of mathematically literate students

The *Investigations* curriculum embodies an approach radically different from the traditional textbook-based curriculum. At each grade level, it consists of a set of separate units, each offering 2–4 weeks of work. These units of study are presented through investigations that involve students in the exploration of major mathematical ideas.

Approaching the mathematics content through investigations helps students develop flexibility and confidence in approaching problems, fluency in using mathematical skills and tools to solve problems, and proficiency in evaluating their solutions. Students also build a repertoire of ways to communicate about their mathematical thinking, while their enjoyment and appreciation of mathematics grows.

The investigations are carefully designed to invite all students into mathematics—girls and boys, diverse cultural, ethnic, and language groups, and students with different strengths and interests. Problem contexts often call on students to share experiences from their family, culture, or community. The curriculum eliminates barriers—such as work in isolation from peers, or emphasis on speed and memorization—that exclude some students from participating successfully in mathematics. The following aspects of the curriculum ensure that all students are included in significant mathematics learning:

- Students spend time exploring problems in depth.
- They find more than one solution to many of the problems they work on.
- They invent their own strategies and approaches, rather than relying on memorized procedures.
- They choose from a variety of concrete materials and appropriate technology, including calculators, as a natural part of their everyday mathematical work.
- They express their mathematical thinking through drawing, writing, and talking.
- They work in a variety of groupings—as a whole class, individually, in pairs, and in small groups.
- They move around the classroom as they explore the mathematics in their environment and talk with their peers.

While reading and other language activities are typically given a great deal of time and emphasis in elementary classrooms, mathematics often does not get the time it needs. If students are to experience mathematics in depth, they must have enough time to become engaged in real mathematical problems. We believe that a minimum of five hours of mathematics classroom time a week—about an hour a day—is critical at the elementary level. The plan and pacing of the *Investigations* curriculum is based on that belief.

For further information about the pedagogy and principles that underlie these investigations, see the Teacher Notes throughout the units and the following books:

- *Implementing the* Investigations in Number, Data, and Space™ *Curriculum*
- *Beyond Arithmetic*

HOW TO USE THIS BOOK

The *Investigations* curriculum is presented through a series of teacher books, one for each unit of study. These books not only provide a complete mathematics curriculum for your students, they offer materials to support your own professional development. You, the teacher, are the person who will make this curriculum come alive in the classroom; the book for each unit is your main support system.

While reproducible resources for students are provided, the curriculum does not include student books. Students work actively with objects and experiences in their own environment and with a variety of manipulative materials and technology, rather than with workbooks and worksheets filled with problems. We also make extensive use of the overhead projector as a way to present problems, to focus group discussion, and to help students share ideas and strategies. If an overhead projector is available, we urge you to try it as suggested in the investigations.

Ultimately, every teacher will use these investigations in ways that make sense for his or her particular style, the particular group of students, and the constraints and supports of a particular school environment. We have tried to provide with each unit the best information and guidance for a wide variety of situations, drawn from our collaborations with many teachers and students over many years. Our goal in this book is to help you, as a professional educator, implement this mathematics curriculum in a way that will give all your students access to mathematical power.

Investigation Format

The opening two pages of each investigation help you get ready for the student work that follows. Here you will read:

What Happens—a synopsis of each session or block of sessions.

Mathematical Emphasis—the most important ideas and processes students will encounter in this investigation.

What to Plan Ahead of Time—materials to gather, student sheets to duplicate, transparencies to make, and anything else you need to do before starting.

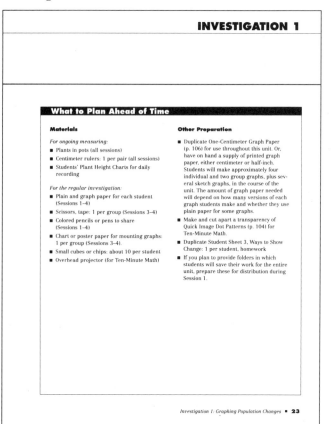

2 ▪ *Changes Over Time*

Sessions Within an investigation, the activities are organized by class session, a session being a one-hour math class. Sessions are numbered consecutively through an investigation. Often several sessions are grouped together, presenting a block of activities with a single major focus.

When you find a block of sessions presented together—for example, Sessions 1, 2, and 3—read through the entire block first to understand the overall flow and sequence of the activities. Make some preliminary decisions about how you will divide the activities into three sessions for your class, based on what you know about your students. You may need to modify your initial plans as you progress through the activities, and you may want to make notes in the margins of the pages as reminders for the next time you use the unit.

Be sure to read the Session Follow-Up section at the end of the session block to see what homework assignments and extensions are suggested as you make your initial plans.

While you may be used to a curriculum that tells you exactly what each class session should cover, we have found that the teacher is in a better position to make these decisions. Each unit is flexible and may be handled somewhat differently by every teacher. While we provide guidance for how many sessions a particular group of activities is likely to need, we want you to be active in determining an appropriate pace and the best transition points for your class.

Ten-Minute Math At the beginning of some sessions, you will find Ten-Minute Math activities. These are designed to be used in tandem with the investigations, but not during the math hour. Rather, we hope you will do them whenever you have a spare 10 minutes—maybe before lunch or recess, or at the end of the day.

Ten-Minute Math offers practice in key concepts, but not always those being covered in the unit. For example, in a unit on using data, Ten-Minute Math might revisit geometric activities done earlier in the year. Complete directions for the suggested activities are included at the end of each unit. A compilation of Ten-Minute Math activities is also available as a separate book.

Sessions 1 and 2

Graphing Home Population Over Time

Materials

- Plain paper and graph paper for each student
- Colored pencils or pens
- Small cubes or chips (10 per group)
- Student Sheet 3 (1 per student, homework)

What Happens

Students make graphs to show the changing population in their homes over a typical day. They work in groups to interpret each other's graphs and plan how they will revise the graphs. Their work focuses on:

- making graphs that show changes over time
- inventing representations
- interpreting different kinds of graphs

Note: Students continue to care for and measure their plants and record heights daily on their copies of Student Sheet 2, Plant Height Chart.

Activity

Graphing Population at Home

Introducing the Unit Explain that during this unit, you will be looking at how things change over time. Tell your students that they will first look at how the number of people in their home changes over a day and then at how the height of their bean plant changes over a few weeks. They will make graphs to keep track for themselves and to communicate with other people what changes they observe.

The First Graph Ask students to take out the information they collected about the number of people at home at different times during a day.

Today you are going to use this information to make a graph that shows how many people are at your home at different times during the day and night. Your graph should be clear enough that someone else can tell how many people are at home at any given time and when people go in and out.

Make both graph paper and plain paper available. Encourage students to make their graphs in any way they want. Some will choose to write the exact times people come and go. Others will not. Do not insist that students write the times at regular hour or half-hour intervals. It is important that they represent the information in a way that makes sense to them.

24 ■ *Investigation 1: Graphing Population Changes*

Activities The activities include pair and small-group work, individual tasks, and whole-class discussions. In any case, students are seated together, talking and sharing ideas during all work times. Students most often work cooperatively, although each student may record work individually.

Choice Time In some units, some sessions are structured with activity choices. In these cases, students may work simultaneously on different activities focused on the same mathematical ideas. Students choose which activities they want to do, and they cycle through them.

You will need to decide how to set up and introduce these activities and how to let students make their choices. Some teachers present them as station activities, in different parts of the room. Some list the choices on the board as reminders or have students keep their own lists.

Excursions Some of the investigations in this unit include *excursions*—activities that could be omitted without harming the integrity of the unit. This is one way of dealing with the overabundance of fascinating mathematics to be studied—much more than a class has time to explore in any one year. Excursions give you the flexibility to make different choices from year to year. You might do the excursions in *Changes Over Time* this year, but another year, try the excursions in another unit.

Tips for the Linguistically Diverse Classroom
At strategic points in each unit, you will find concrete suggestions for simple modifications of the teaching strategies to encourage the participation of all students. Many of these tips offer alternative ways to elicit critical thinking from students at varying levels of English proficiency, as well as from other students who find it difficult to verbalize their thinking.

The tips are supported by suggestions for specific vocabulary work to help ensure that all students can participate fully in the investigations. The Preview for the Linguistically Diverse Classroom (p. 12) lists important words that are assumed as part of the working vocabulary of the unit. Second-language learners will need to become familiar with these words in order to understand the problems and activities they will be doing. These terms can be incorporated into students' second-language work before or during the unit. Activities that can be used to present the words and make them comprehensible are found in the appendix, Vocabulary Support for Second-Language Learners (p. 86).

In addition, ideas for making connections to students' language and cultures, included on the Preview page, help the class explore the unit's concepts from a multicultural perspective.

Session Follow-Up
Homework Homework is not given daily for its own sake, but periodically as it makes sense to have follow-up work at home. Homework may be used for (1) review and practice of work done in class; (2) preparation for activities coming up—for example, collecting data for a class project; or (3) involving and informing family members.

Some units in the *Investigations* curriculum have more homework than others, simply because it makes sense for the mathematics that's going on. Other units rely on manipulatives that most students won't have at home, making homework difficult. In any case, homework should always be directly connected to the investigations in the unit, or to work in previous units—never sheets of problems just to keep students busy.

Extensions These follow-up activities are opportunities for some or all students to explore a topic in greater depth or in a different context. They are not designed for "fast" students; mathematics is a multifaceted discipline, and different students will want to go further in different investigations. Look for and encourage the sparks of interest and enthusiasm you see in your students, and use the extensions to help them pursue these interests.

Family Letter A letter that you can send home to students' families is included with the blackline masters for each unit. We want families to be informed about the mathematics work in your classroom; they should be encouraged to participate in and support their children's work. A reminder to send home the letter appears in one of the early investigations. (These letters are also available separately in the following languages: Spanish, Vietnamese, Cantonese, Hmong, and Cambodian.)

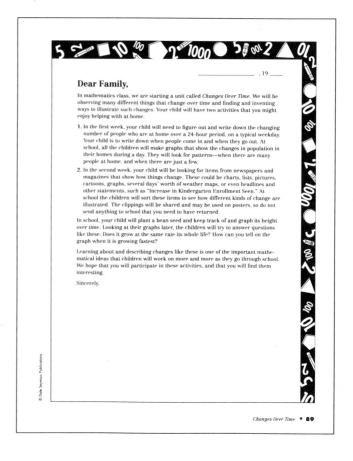

Materials

A complete list of the materials needed for the unit is found on p. 10. Some of these materials are available in a kit for the *Investigations* grade 4 curriculum. Individual items can also be purchased as needed from school supply stores and dealers.

In an active mathematics classroom, certain basic materials should be available at all times: interlocking cubes, pencils, unlined paper, graph paper, calculators, things to count with, and measuring tools. Some activities in this curriculum require scissors and glue sticks or tape. Stick-on notes and large paper are also useful materials throughout.

So that students can independently get what they need at any time, they should know where these materials are kept, how they are stored, and how they are to be returned to the storage area. For example, interlocking cubes are best stored in towers of ten; then, whatever the activity, they should be returned to storage in groups of ten at the end of the hour. You'll find that establishing such routines at the beginning of the year is well worth the time and effort.

Student Sheets and Teaching Resources

Reproducible pages to help you teach the unit are found at the end of this book. These include masters for making overhead transparencies and other teaching tools, as well as student recording sheets.

Many of the field-test teachers requested more sheets to help students record their work, and we have tried to be responsive to this need. At the same time, we think it's important that students find their own ways of organizing and recording their work. They need to learn how to explain their thinking with both drawings and written words, and how to organize their results so someone else can understand them.

To ensure that students get a chance to learn how to represent and organize their own work, we deliberately do not provide student sheets for every activity. We recommend that your students keep a mathematics notebook or folder so that their work, whether on reproducible sheets or their own paper, is always available to them for reference.

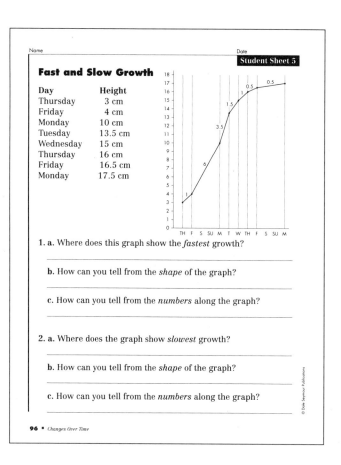

Help for You, the Teacher

Because we believe strongly that a new curriculum must help teachers think in new ways about mathematics and about their students' mathematical thinking processes, we have included a great deal of material to help you learn more about both.

About the Mathematics in This Unit This introductory section (p. 11) summarizes for you the critical information about the mathematics you will be teaching. This will be particularly valuable to teachers who are accustomed to a traditional textbook-based curriculum.

Teacher Notes These reference notes provide practical information about the mathematics you are teaching and about our experience with how students learn. Many of the notes were written in response to actual questions from teachers, or to discuss important things we saw happening in the field-test classrooms. Some teachers like to read them all before starting the unit, then review them as they come up in particular investigations.

Dialogue Boxes Sample dialogues demonstrate how students typically express their mathematical ideas, what issues and confusions arise in their thinking, and how some teachers have guided class discussions.

These dialogues are based on the extensive classroom testing of this curriculum; many are word-for-word transcriptions of recorded class discussions. They are not always easy reading; sometimes it may take some effort to unravel what the students are trying to say. But this is the value of these dialogues; they offer good clues to how your students may develop and express their approaches and strategies, helping you prepare for your own class discussions.

Where to Start You may not have time to read everything the first time you use this unit. As a first-time user, you will likely focus on understanding the activities and working them out with your students. Read completely through each investigation before starting to present it.

When you next teach this same unit, you can begin to read more of the background. Each time you present this unit, you will learn more about how your students understand the mathematical ideas. The first-time user of *Changes Over Time* should read the following:

- About the Mathematics in This Unit (p. 11)
- Teacher Note: Varieties of Students' Graphs (p. 29)
- Teacher Note: Students' Line Graphs (p. 64)
- Teacher Note: Assessment: Telling Stories from Graphs (p. 72)

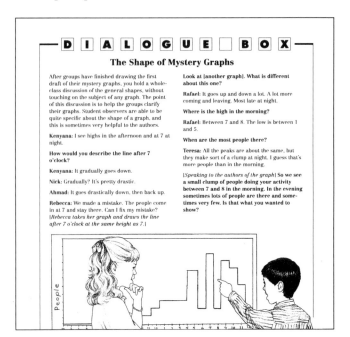

ABOUT ASSESSMENT

Teacher Checkpoints As a teacher of the *Investigations* curriculum, you observe students daily, listen to their discussions, look carefully at their work, and use this information to guide your teaching. We have designated Teacher Checkpoints as natural times to get an overall sense of how your class is doing in the unit.

The Teacher Checkpoints provide a time for you to pause and reflect on your teaching plan while observing students at work in an activity. These sections offer tips on what you should be looking for and how you might adjust your pacing. Are most students fluent with strategies for solving a particular kind of problem? Are they just starting to formulate good strategies? Or are they still struggling with how to start?

Depending on what you see as the students work, you may want to spend more time on similar problems, change some of the problems to use smaller numbers, move quickly to more challenging material, modify subsequent activities for some students, work on particular ideas with a small group, or pair students who have good strategies with those who are having more difficulty.

In *Changes Over Time* you will find two Teacher Checkpoints:

> Do Their Graphs Work? (p. 28)
> Our Growing Plants (p. 62)

Embedded Assessment Activities Use the built-in assessments included in this unit to help you examine the work of individual students, figure out what it means, and provide feedback. From the students' point of view, the activities you will be using for assessment are no different from any others; they don't look or feel like traditional tests.

These activities sometimes involve writing and reflecting, at other times a brief interaction between student and teacher, and in still other instances the creation and explanation of a product.

In *Changes Over Time* you will find assessment activities in the first and third investigations:

> Revising Individual Graphs (p. 32)
> Telling Stories from Graphs (p. 71)

Teachers find the hardest part of the assessment to be interpreting their students' work. If you have used a process approach to teaching writing, you will find our mathematics approach familiar. To help with interpretation, we provide guidelines and questions to ask about the students' work. In some cases we include a Teacher Note with specific examples of student work and a commentary on what it indicates. This framework can help you determine how your students are progressing.

As you evaluate students' work, it's important to remember that you're looking for much more than the "right answer." You'll want to know what their strategies are for solving the problem, how well these strategies work, whether they can keep track of and logically organize an approach to the problem, and how they make use of representations and tools to solve the problem.

Ongoing Assessment Good assessment of student work involves a combination of approaches. Some of the things you might do on an ongoing basis include the following:

- **Observation** Circulate around the room to observe students as they work. Watch for the development of their mathematical strategies, and listen to their discussions of mathematical ideas.

- **Portfolios** Ask students to document their work, in journals, notebooks, or portfolios. Periodically review this work to see how their mathematical thinking and writing are changing. Some teachers have students keep a notebook or folder for each unit, while others prefer one mathematics notebook or a portfolio of selected work for the entire year. Take time at the end of each unit for students to choose work for their portfolios. You might also have them write about what they've learned in the unit.

Changes Over Time

OVERVIEW

Content of This Unit Students explore the ways things change over time and look at ways to represent these changes. Looking at change is a recent development in elementary mathematics. Typically students' graphing experiences focus on comparing things that take place at the same time, for example, choices of TV shows or sizes of families. This unit builds on previous graphing experiences as students make representations to show how one thing changes over time. After finding their own ways to show the changing population of their homes during a day, students make charts and line graphs to show changing speed, changing heights of plants over several weeks, and changing population. They interpret the shapes of the curves in line graphs.

Connections with Other Units If you are doing the full-year *Investigations* curriculum in the suggested sequence for grade 4, this is the eighth of 11 units. Students will have made line plots and looked for typical data in the Statistics unit, *The Shape of the Data*. In the later 2-D Geometry unit, *Sunken Ships and Grid Patterns,* students will have experience with plotting points in four quadrants. In the grade 3 Changes unit, students investigated and graphed changes along a number line. In the grade 5 Changes unit, they will graph things that change in regular ways, such as the areas of squares, and look at how they change as they grow larger.

This unit can be used successfully at grades 4, 5, or 6, depending on the previous experience and needs of your students.

Investigations Curriculum ■ Suggested Grade 4 Sequence

Mathematical Thinking at Grade 4 (Introduction)

Arrays and Shares (Multiplication and Division)

Seeing Solids and Silhouettes (3-D Geometry)

Landmarks in the Thousands (The Number System)

Different Shapes, Equal Pieces (Fractions)

The Shape of the Data (Statistics)

Money, Miles, and Large Numbers (Addition and Subtraction)

▶ *Changes Over Time* (Graphs)

Packages and Groups (Multiplication and Division)

Sunken Ships and Grid Patterns (2-D Geometry)

Three out of Four Like Spaghetti (Data and Fractions)

Unit Preparation: Growing Plants to Graph

About a week and a half before starting this unit, students need to start growing plants for use in Investigation 3. Allow for three short preparation sessions, several days apart, to be done while you are working on another mathematics unit. See the Scheduling Notes for This Unit (p. 14) for more information.

Class Sessions	Activities	Pacing
Preparation Session 1 SPROUTING THE SEEDS	Sharing Ideas About Plant Growth Getting Seeds Ready to Sprout	0.5 hr
Preparation Session 2 PLANTING THE SEEDS	Planting Sprouted Seeds ■ Homework	0.5 hr
Preparation Session 3 MEASURING THE PLANTS	Starting the Plant Height Chart	0.5 hr

Investigation 1 • Graphing Population Changes

Class Sessions	Activities	Pacing	Ten-Minute Math
Sessions 1 and 2 GRAPHING HOME POPULATION OVER TIME	Graphing Population at Home Interpreting Each Other's Graphs ■ Teacher Checkpoint: Do Their Graphs Work? ■ Homework	2 hrs	Quick Images
Sessions 3 and 4 MAKING GROUP GRAPHS	Combining Data on One Graph ■ Assessment: Revising Individual Graphs Assembling Posters Looking at the Overall Shape ■ Homework	2 hrs	
Sessions 5 and 6 (Excursion)* INS AND OUTS NUMBER PROBLEMS	Ins and Outs Problems Solving Backward Problems Writing Story Problems Solving Our Story Problems ■ Extension	2 hrs	

* Excursions can be omitted without harming the integrity or continuity of the unit, but offer good mathematical work if you have time to include them.

Investigation 2 • Ways to Show Change Over Time

Class Sessions	Activities	Pacing	Ten-Minute Math
Sessions 1 and 2 SHOWING CHANGE OVER TIME	Looking at Ways to Show Change Showing Change Our Own Ways ■ Homework ■ Extension	2 hrs	Quick Images

Investigation 3 • Telling Stories from Line Graphs

Class Sessions	Activities	Pacing
Session 1 MAKING GRAPH SKETCHES	Discussing Overall Shapes of Graphs Describing a Marathon Graphically Interpreting the Graphs Another Graph of Motion	1 hr

Continued on next page

Investigation 3 • Telling Stories from Line Graphs (continued)

Class Sessions	Activities	Pacing	Ten-Minute Math
Session 2 GRAPHING AND PREDICTING PLANT GROWTH	Graphing Our Plant Heights Predicting Future Growth Faster and Slower Growth ■ Teacher Checkpoint: Our Growing Plants ■ Homework	1 hr	Broken Calculator Problems
Session 3 USING LINE GRAPHS TO COMPARE GROWTH	Combining Graphs of Different Plants ■ Homework	1 hr	
Session 4 GRAPHS, STORIES, AND NUMBER SEQUENCES	Matching Numbers, Stories, and Graphs	1 hr	
Session 5 INTERPRETING GRAPHS	■ Assessment: Telling Stories from Graphs	1 hr	
Sessions 6 and 7 MYSTERY GRAPHS	Making Graphs of Changing Population Discussing the Shapes of the Graphs Matching Graphs to Activities Revising the Mystery Graphs ■ Extensions	2 hrs	

MATERIALS LIST

Following are the basic materials needed for the activities in this unit. Items marked with an asterisk are provided with the Investigations Materials Kit for grade 4.

* Small cubes or chips: 10 per student
* Graph paper
   Plain unlined paper
   11" × 17" (or larger) paper
   Chart or poster paper
   Colored pencils or pens
   Scissors
   Glue or tape
   Overhead projector

The following materials are provided at the end of this book as blackline masters. They are also available in classroom sets.

Family Letter (p. 90)

Student Sheets 1–9 (pp. 91–102)

Teaching Resources
   Motion Graph (p. 103)
   Quick Image Dot Patterns (p. 104)
   Quick Image Geometric Designs (p. 105)
   One-Centimeter Graph Paper (p. 106)

For the Preparation Sessions

* Lima bean seeds or dried lima beans for planting, 3 seeds per pair
* Centimeter rulers, 1 per pair
   Paper towels
   Transparent cups or jars for sprouting the seeds, 1 per pair
   Labels or masking tape
   Plastic wrap, 1 roll
   Elastic bands
   Plant pots (peat moss pots work well) or small milk cartons with a diameter large enough for a ruler width, 1 per pair plus 5–6 extras
   Potting soil, about 5 pounds
   Sticks and string or twist ties for holding plants straight (optional)
   Plant fertilizer (optional)

ABOUT THE MATHEMATICS IN THIS UNIT

Following are some of the major mathematical ideas students meet in this unit of data study:

Understanding how changes and total are related. A number can describe a change in the size or amount of something at a given time (for example, we may watch 3 cars enter the parking lot), or it can describe the total size or amount accumulated over a period of time (for example, by now there are 25 cars in the parking lot). A critical mathematical issue is to relate numbers that describe changes to numbers that describe the resulting size or amount. (There are 7 cars at the beginning. Then 5 go out and 2 come in, so then there are 4.)

Understanding the difference between continuous and discrete changes. Things that change in discrete units, such as the number of people in a room, can be described with whole numbers and can usually be counted. Students explore this type of change in Investigation 1, Graphing Population Changes. Things that change on a continuum, such as the height of a plant, are described with whole numbers plus all the numbers in between, and must be measured. Students explore continuous change in Investigation 3, Telling Stories from Line Graphs.

Describing data. In this unit, students integrate three types of observations to describe data: *quantitative* observations, such as heights of plants or numbers of people; *qualitative* observations, such as noticing when things are growing faster or slowing down; and *graphical* observations, such as noticing a line that goes up quickly.

Getting an overall sense of change from a graph. We can get information about change from the shape of a graph without looking at the actual numbers. Is the graph going up or down? That tells us at a glance if the plant is increasing or decreasing in height. Is the graph a straight line or a curve? That tells us if the plant is growing at a steady rate or a changing speed. How steep is its slope? That tells us how rapid the rate of growth is.

Mathematical Emphasis At the beginning of each investigation, the Mathematical Emphasis section tells you what is most important for students to learn about during that investigation. Many of these mathematical understandings and processes are difficult and complex. Students gradually learn more and more about each idea over many years of schooling. Individual students will begin and end the unit with different levels of knowledge and skill, but all students will gain greater knowledge about how to describe the way something changes over time with numbers, graphs, and qualitative descriptions.

PREVIEW FOR THE LINGUISTICALLY DIVERSE CLASSROOM

In the *Investigations* curriculum, mathematical vocabulary is introduced naturally during the activities. We don't ask students to learn definitions of new terms; rather, they come to understand such words as *factor* or *area* or *symmetry* by hearing them used frequently in discussion as they investigate new concepts. This approach is compatible with current theories of second-language acquisition, which emphasize the use of new vocabulary in meaningful contexts while students are actively involved with objects, pictures, and physical movement.

Listed below are some key words used in this unit that will not be new to most English speakers at this age level, but may be unfamiliar to students with limited English proficiency. You will want to spend additional time working on these words with your students who are learning English. If your students are working with a second-language teacher, you might enlist your colleague's aid in familiarizing students with these words, before and during this unit. In the classroom, look for opportunities for students to hear and use these words. Activities you can use to present the words are given in the appendix, Vocabulary Support for Second-Language Learners (p. 86).

in, out In order to make graphs of the changing population of their homes over a 24-hour period, students gather information about when people typically come *in* and go *out*. Later they solve problems by examining different series of ins and outs.

change, size, age, color, amount Students study indications of change in size, color, and so on, in preparation for making graphs that show change over time.

runner, wheelchair racer, marathon, speed, slow, fast, quickly, pace, gradually, steadily Students graph a wheelchair racer's progress in a marathon, showing the varying speeds and changes of speed during the race.

Multicultural Extensions for All Students

Whenever possible, encourage students to share words, objects, customs, or any aspects of daily life from their own cultures and backgrounds that are relevant to the activities in this unit. For example, when students are finding ways to change in Investigation 2, they can bring in or draw examples that relate specifically to their culture or country of origin. Someone might bring pictures of their family both in another country and here. Or a student might contribute a specific item such as a piñata, before and after.

Investigations

UNIT PREPARATION

Growing Plants to Graph

Scheduling Notes for This Unit

As part of their study of change in this unit, students plant beans and follow the growth of their plants for four weeks. For this reason, you will want to avoid starting the unit within four weeks of a vacation period. It is fine to leave the plants unattended for two- or three-day weekends. The general schedule is as follows:

Getting Ready About 10–12 days before you plan to begin Investigation 1, soak the beans in water overnight. Proceed the next day with Preparation Session 1.

Preparation Session 1: Sprouting the Seeds
Students share their ideas about how plants grow and then prepare their seeds for sprouting. After this session, wait several days for the seeds to sprout. This will take 3–5 days. When most seeds have sprouted, hold Preparation Session 2.

Preparation Session 2: Planting the Seeds
Students plant their sprouted seeds in soil. Then you will need to wait 4–6 days while the plants begin to grow. At this time, you will also send home guidelines for collecting data for a "home population graph," which students will need as they begin Investigation 1. When some of the plant stems have grown to 1 cm tall, you are ready for Preparation Session 3.

Preparation Session 3: Measuring the Plants
Students discuss guidelines for measuring plants. They begin to keep track of their plants' heights in a chart. Preparation is now complete; proceed the next day with Investigation 1.

Investigations 1, 2, and 3 Over a period of 13–15 class sessions, students will care for and measure their plants and record stem heights daily, while also doing other activities related to change over time. During Investigation 3, students will be graphing their plants' growth.

UNIT PREPARATION

What to Plan Ahead of Time

Materials

- Lima bean seeds or dried lima beans for planting, 3 per pair plus extras (Session 1)
- Paper towels, 2 per pair (Session 1)
- Transparent plastic cups or jars, 1 per pair (Session 1)
- Plastic wrap: enough to cover each cup (Session 1)
- Elastic bands: 1 per pair (Session 1)
- Marking pens (Sessions 1–2)
- Labels or masking tape for marking plant containers (Sessions 1–2)
- Plant pots (peat moss pots work well) or small milk cartons with a diameter large enough for a ruler to stand in on end, 1 per pair plus 5–6 extras (Session 2)
- Potting soil, about 5 pounds (Session 2)
- Plant fertilizer (Session 2, optional)
- Centimeter rulers: 1 per pair (Session 3)
- Sticks and string or twist ties for holding plants straight (Session 3, optional)

Other Preparation

- 10–12 days before starting Investigation 1, put seeds into a cup or bowl and cover them with water. Allow them to soak overnight.
- Duplicate student sheets and teaching resources (located at the end of this unit) as follows:

  Student Sheet, 1 Getting Data About Home Population: 1 per student

  Family Letter (p. 90): 1 per student. Remember to sign the letter before copying.

  Student Sheet 2, Plant Height Chart: 1 per student

Preparation Session 1

Sprouting the Seeds

Materials

- Lima bean seeds, soaked (3 per pair plus extras)
- Paper towels (2 per pair)
- Transparent plastic cups or jars (1 per pair)
- Plastic wrap (enough to cover each cup)
- Elastic bands (1 per pair)
- Marking pens
- Labels or masking tape

What Happens

Students share their ideas about how plants grow and then prepare their seeds for sprouting. After this session, wait several days for the seeds to sprout. This will take 3–5 days. When most seeds have sprouted, hold Preparation Session 2.

Activity

Sharing Ideas About Plant Growth

Show students the seeds that have been soaking. Explain that they will be taking care of some seeds, first sprouting them and then planting them in order to discover how they change over time. Ask students to share their ideas about how the plants will change.

What do you think will happen first? After that? How much do you think the plants will grow each day? Do you think that the plants will grow the same or a different amount each day? Do you think they will grow faster when they start or after a week or two?

Students will probably have different ideas about the answers to these questions. Allow them to share their ideas without commenting on their correctness at this time.

Activity

Getting Seeds Ready to Sprout

Demonstrate the following steps. Students who are absent for this session may use the seeds you prepare.

- Wet two pieces of paper towel and wring out the excess water.
- Fold the paper towels into a strip about as wide as the height of the cup, and curve it around the inside of the cup.
- Place the seeds between the cup and the paper towel so that the paper towel holds them in place.
- Add water to touch the paper towel but not cover the seeds.
- Put plastic wrap over the top of the cup to limit the evaporation.
- Label the container with students' names and put it in a dark place until the seeds sprout.

Pairs of students select three of the soaked seeds and follow the procedure you demonstrated. Establish a prescribed time, approximately the same time each day, for students to check on their seeds and add water if necessary. If a seed becomes moldy, students should remove it and change the paper towel. Remind your students to keep the water level so that it touches the towel but does not cover the seeds. This monitoring does not need to be part of mathematics class, but it must not be forgotten.

Tell your students that after a few days, when the roots of their plants have grown some, they will plant them in dirt. When most of the students' seeds have roots, follow the instructions in Preparation Session 2, Planting the Seeds.

Sprouting the Seeds

Preparation Session 2

Planting the Seeds

Materials

- Sprouted seeds in cups
- Plant pots (1 per student)
- Potting soil (about 5 pounds)
- Labels or masking tape
- Marking pens
- Plant fertilizer (optional)
- Student Sheet 1 (1 per student)
- Family letter (1 per student)

What Happens

Students plant their seeds in soil. Then you will need to wait 4–6 days while the plants begin to grow. At this time, you will also send home guidelines for collecting data for a "home population graph," which students will need as they begin Investigation 1. When some of the plant stems have grown to 1 cm tall, you are ready for Preparation Session 3.

Activity

Planting Sprouted Seeds

Students, working with their partners, choose two healthy seeds that have roots. If they are using milk cartons as planting pots, be sure that they put a small drainage hole in the bottom. For each seed, they fill a pot almost to the top with soil and soak the soil with water. Then, with their finger, they make a hole in the soil deep enough to accommodate the seed. They gently place the sprouted seed, root down, in the hole. They pack the soil around the seed loosely, and put a shallow layer of soil over it. They label the plant pots with their names and perhaps a name for each plant.

The plants should be allowed to dry out between waterings and should be kept in a warm, brightly lit window, if possible. Continue to have students tend plants daily.

We recommend that students work in pairs with only one plant. They can begin by keeping both plants until one germinates. They keep the plant that germinates first and take daily measurements of its height. They put aside the other planted seed, or if it has also germinated, they can give it to students who have no germinated plant. Collect a few of the extra planted seeds to care for in case something happens to some students' plants. You will need to keep track of the heights of these plants once they germinate so you can pass on the information to students who assume their care.

Once the stems of the plants start to grow, your students may want to add a plant fertilizer to the water. After four to six days, when some of the stems have grown to about 1 centimeter, the students will start to keep track of their plants' changing heights. At this point, see Preparation Session 3, Measuring the Plants.

Preparation Session 2 Follow-Up

🏠 **Homework**

One or two days before beginning Investigation 1, distribute Student Sheet 1, Getting Data About Home Population. Tell students that they will be making graphs that show how the number of people in their homes increases and decreases during one day. Ask them to bring to class the following information: how many people are at their home during different periods of time during a *typical school day,* from midnight until midnight, and the typical times people leave and return. Suggest that they talk to an adult to help determine the usual times that people in their household go away from home (not just outside) and return.

❖ **Tip for the Linguistically Diverse Classroom** To make sure students who are not fluent in English understand the instructions for Student Sheet 1, present an example using pantomime, simple drawings, and other visual aids. For example, you might use a calendar and clock, a drawing of a house and stick figures, and pantomime of comings and goings to show the population rise and fall in a household during a day.

As you make this assignment, there may be some discussion about what counts as a student's home. For example, "Do I count my grandmother who lives in a separate apartment upstairs?" "I live sometimes with my mother and sometimes with my father. Which should I do? Can I mix them?" In addition, a household of only two people might not be interesting to graph unless other people visit regularly, and a large household might be difficult for some students to keep track of. You do not have to make rules for the class as a whole; just help the students decide for themselves.

Send the family letter home with Student Sheet 1.

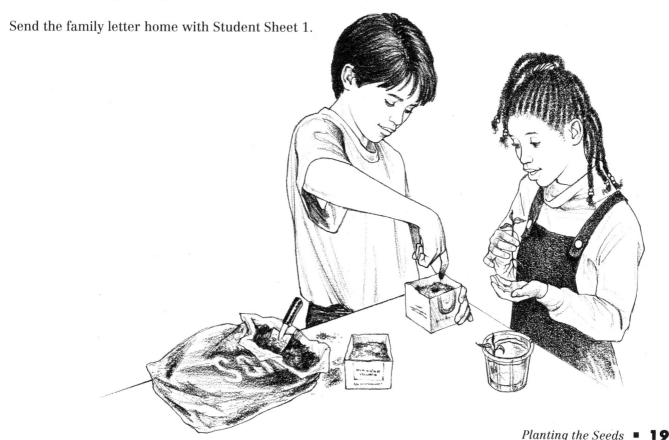

Preparation Session 3

Measuring the Plants

Materials

- Plants in pots
- Centimeter rulers (1 per pair)
- Student Sheet 2 (1 per student)
- Sticks and string (optional)

What Happens

Students discuss guidelines for measuring plants. They begin to keep track of their plants' heights in a chart. (Preparation is now complete; proceed the next day with Investigation 1.) Their work focuses on:

- measuring to the nearest half centimeter
- keeping track of daily measurements in a chart

Activity

Starting the Plant Height Chart

As soon as the stem of one or more students' plants is visible, hand out Student Sheet 2, Plant Height Chart, to everyone. Although student pairs will share plants, each student will keep his or her own recording chart. All students begin collecting data at this point. Those whose plants have no visible growth will write 0 cm on their charts.

Use one of the plants whose stem has been growing to demonstrate how to measure. Place a centimeter ruler on the top of the soil. Some rulers have a little extra space at the end before the zero point; in this case, remind students to dig the end of the ruler into the dirt so that the zero mark is just at the surface. Measure to the top of the stem, to the nearest 0.5 centimeter, and point out where this measurement should be recorded on the student sheet. (We recommend using centimeters to avoid dealing with quarter and eighth inches.) Tell your students to measure as carefully as possible to the nearest half centimeter.

Students should also be careful to keep the soil around the base of the stem level. If the stem starts to lean to one side, they should gently straighten it before taking their measurements.

Help students to establish the routine of recording heights daily, initially just on their charts, and then later, in Investigation 3, on their graphs as well.

Unit Preparation: Growing Plants to Graph

Note: Students should not graph their plant heights yet. They will begin this in Investigation 3, Session 2, after some discussion of a sample graph. Once students have started measuring their plants, they should continue measuring them even if the stem breaks, and once they start graphing a certain plant, they should continue with that plant. It is helpful for students to see that a graph of height can go down, and to explain what might cause a sudden or gradual drop in height. If none of the students' plants break, you might break off a bit of one of your plants and keep measuring and graphing it to show the students.

What to Measure? As their plants gain height, students may be unclear about what to measure. In general, they should measure to the highest point on the plant. Usually this means that they will be measuring the height of the main stem. Sometimes, however, a peripheral leaf grows higher than the main stem. In that case, they will measure to the top of the leaf's stem.

It is important to be consistent about what is measured, but some amount of inconsistency is inevitable. It is best to attend to any ambiguities at the time they arise.

Reminder: Once students have begun to measure their plants, you can begin Investigation 1. Be sure students have taken home Student Sheet 1, Getting Data About Home Population, along with the family letter, before you begin the investigation (see p. 19 for a description of the home population data collection).

INVESTIGATION 1

Graphing Population Changes

What Happens

Sessions 1 and 2: Graphing Home Population Over Time Students make graphs to show the changing population in their homes over a typical day. They work in groups to interpret each other's graphs and plan how they will revise the graphs.

Sessions 3 and 4: Making Group Graphs Students work in groups of four, each group making one graph that shows the changes of population in their four homes combined. Each group makes a poster to display the individual and group graphs they have made.

Sessions 5 and 6 (Excursion): Ins and Outs Number Problems Students solve problems that involve finding a missing number in someone's report of the ins and outs from his or her home. Then students write fantasy story problems about people and other creatures coming and going, and trade problems to solve them. Since there is a lot of writing, rewriting, and reading involved, you might do this activity in writing time.

Mathematical Emphasis

- Making graphs that show change over time
- Inventing representations of data
- Interpreting different kinds of graphs
- Developing a scale that includes all the data
- Deciding how to group data
- Establishing conventions for consistency
- Understanding how changes and total are related
- Developing strategies for solving missing information problems when the information is missing from the beginning, middle, or end (Excursion)
- Writing missing information problems (Excursion)

INVESTIGATION 1

What to Plan Ahead of Time

Materials

For ongoing measuring:
- Plants in pots (all sessions)
- Centimeter rulers: 1 per pair (all sessions)
- Students' Plant Height Charts for daily recording

For the regular investigation:
- Plain and graph paper for each student (Sessions 1–4)
- Scissors, tape: 1 per group (Sessions 3–4)
- Colored pencils or pens to share (Sessions 1–4)
- Chart or poster paper for mounting graphs: 1 per group (Sessions 3–4).
- Small cubes or chips: about 10 per student
- Overhead projector (for Ten-Minute Math)

Other Preparation

- Duplicate One-Centimeter Graph Paper (p. 106) for use throughout this unit. Or, have on hand a supply of printed graph paper, either centimeter or half-inch. Students will make approximately four individual and two group graphs, plus several sketch graphs, in the course of the unit. The amount of graph paper needed will depend on how many versions of each graph students make and whether they use plain paper for some graphs.
- Make and cut apart a transparency of Quick Image Dot Patterns (p. 104) for Ten-Minute Math.
- Duplicate Student Sheet 3, Ways to Show Change: 1 per student, homework
- If you plan to provide folders in which students will save their work for the entire unit, prepare these for distribution during Session 1.

Investigation 1: Graphing Population Changes

Sessions 1 and 2

Graphing Home Population Over Time

Materials

- Plain paper and graph paper for each student
- Colored pencils or pens
- Small cubes or chips (10 per group)
- Student Sheet 3 (1 per student, homework)

What Happens

Students make graphs to show the changing population in their homes over a typical day. They work in groups to interpret each other's graphs and plan how they will revise the graphs. Their work focuses on:

- making graphs that show changes over time
- inventing representations
- interpreting different kinds of graphs

Note: Students continue to care for and measure their plants and record heights daily on their copies of Student Sheet 2, Plant Height Chart.

Activity

Graphing Population at Home

Introducing the Unit Explain that during this unit, you will be looking at how things change over time. Tell your students that they will first look at how the number of people in their home changes over a day and then at how the height of their bean plant changes over a few weeks. They will make graphs to keep track for themselves and to communicate with other people what changes they observe.

The First Graph Ask students to take out the information they collected about the number of people at home at different times during a day.

Today you are going to use this information to make a graph that shows how many people are at your home at different times during the day and night. Your graph should be clear enough that someone else can tell how many people are at home at any given time and when people go in and out.

Make both graph paper and plain paper available. Encourage students to make their graphs in any way they want. Some will choose to write the exact times people come and go. Others will not. Do not insist that students write the times at regular hour or half-hour intervals. It is important that they represent the information in a way that makes sense to them.

The choices students make can become ideal focal points for discussions about how their representations communicate. In many cases, the experience of making a graph, getting feedback about its clarity, and seeing how others make graphs will cause students to become more accurate representers of information on their own.

As students refine their graphs, discuss how the systems they invent differ from conventional graphs. These discussions should help them become aware that conventional approaches, such as bar graphs with times in equal intervals or use of an empty space to show zero, are often used because they work clearly for many different situations.

By using traditional systems, students gain regularity, but they may lose some information. For example, when they group everyone who is home between 7 and 8 A.M. as being home for the whole hour, they lose the information about exactly when each person leaves.

Some students who have put the information in a list or chart at home will want to start by copying their work to make it neater. Ask them instead to leave it rough and to make some kind of graph to show the information on it. Some charts, of course, can function as graphs. See the **Teacher Note, Varieties of Students' Graphs** (p. 29), for ideas about what to expect in this activity.

Tell students they will have only half an hour to begin making their graphs. Remind them that these are first drafts and need provide only enough information that someone else would be able to look at the graph and tell the story of people coming in and going out all day long.

Emphasize that the purpose is to make graphs that communicate clearly, not to make beautiful displays. Students do not even need to finish the whole 24 hours before going on to the next activity, in which they will read each other's graphs. They will have some time to revise and finish the graphs after that activity or in the next few days.

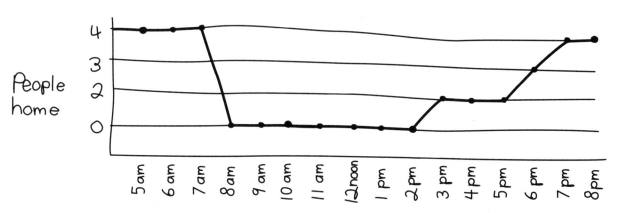

"I didn't put 1 because there was never just one person home."—Rikki

Some students may have difficulty getting started. Remind them of other graphing that they have done—bar graphs they made in earlier grades or the line plots they made if they have done the grade 4 Statistics unit, *The Shape of the Data*. Resist the urge to show them exactly what to do or to insist on any traditional system.

As students work, ask questions that give them a chance to explain what they are doing. Be sure that, while looking at each graph, you can easily tell a story of how many people are home at different times through the day. If something is not clear to you, such as the number of people or the times of day, ask about it and suggest that students find a way to make it clearer. If a format is hard to understand even when the student explains it to you, help the student to plan a new format and start again.

Students are ready to move on to the next activity when they have begun a graph and put some data in it—perhaps from midnight to noon—with a format that clearly shows the changing population.

Activity

Interpreting Each Other's Graphs

As they are ready, each pair of students gets together with another pair they have not been working with to interpret each other's graphs. You can explain this activity to the whole class at once or to groups of four as they become ready. Pass out cubes to each group to represent people, and a piece of paper to represent a home.

Trade graphs with the other pair. Take turns telling the story of each graph to the rest of the group. Pretend the paper is your classmate's home and the cubes are the people. Move the cubes in and out of the home [on and off the paper] as you tell the story shown on your classmate's graph.

Note: Sometimes a student will put some cubes on the paper to represent people in the home at a given hour, and then take off all the cubes and put some back to show a different number of people at the next hour. Instead, encourage students to move the cubes *as if they were the people going in and out*. That is, if someone stays home for several hours, leave that cube untouched while you are making other changes during that time period.

While one student tells the story of another student's graph or chart, the rest of the group, including the author, listens. Authors should avoid explaining their graphs. Let students know that the main purpose of this activity is to help them see what they need to make clearer on their graphs. Authors can make brief notes directly on their graphs to help them remember what to fix.

❖ **Tip for the Linguistically Diverse Classroom** Pair students to tell the stories of each other's graphs so that a fluent English speaker (you or a student) tells the story while a non-fluent speaker moves the figures in and out of the house.

It is important that students realize that if others do not understand their graph, it is not the fault of the readers. It is the author's responsibility to make the graph clear. Observe to see whose graphs were unclear, and ask what they plan to add or change to solve the problem.

Groups that finish early can begin to revise their graphs.

Looking at the Variety of Approaches When all groups have finished explaining each other's graphs and deciding how to improve their own (at least half an hour before the end of Session 2), gather students to look at more of their classmates' graphs to get ideas.

Collect a graph from each student and redistribute them so that no one gets one from his or her foursome. Working in pairs or in fours, students check the graphs to make sure they are clear and to look for features they find especially helpful. Write questions such as these on the board for students to consider:

How does the graph you are looking at show …

    how many people are at home?

    when most people come home?

    times when no one is home?

What makes the graph easy to understand?

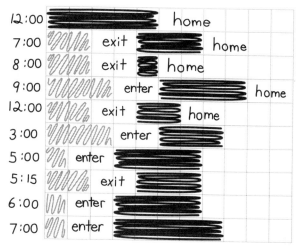

Graphing Home Population Over Time ■ **27**

One teacher gave students stick-on notes so they could put helpful suggestions on the graphs or tell what they especially liked about a particular graph.

If there is time, you can redistribute the graphs again so groups of students can look over another set. Bring students together briefly to share some of their responses.

What ideas have you learned from other people's graphs that you think work well? How are you planning to fix your own graph?

In any time left, students revise their graphs.

Teacher Checkpoint

Do Their Graphs Work?

This is a key checkpoint in this unit. Rather than being taught one way to graph changes, students have been inventing their own types of graphs. Before moving on, be sure that everyone has a good working plan. By the end of Session 2, each student should have a graph that someone else can read and interpret—or at least be able to describe to you a plan for fixing their graph before the next session. You may want to refer again to the **Teacher Note,** Varieties of Students' Graphs, for some perspective on what makes a successful, even though unconventional, student graph.

Sessions 1 and 2 Follow-Up

- Students may finish their graphs at home, if you can be quite sure they will bring them back for the next class session.
- Send home Student Sheet 3, Ways to Show Change, which asks students to collect items from newspapers, magazines, and other sources that show changes over time. Items can include photographs, drawings, cartoons, advertisements, graphs, and so on. Suggest that students ask their families for help, and remind them to ask their families before they cut out anything.

  Make your own collection to supplement the students' for the next investigation.

Varieties of Students' Graphs

> Teacher Note

Unlike conventional graphs, students' invented graphs tend to reflect the information they have without following all the rules of a system. You may see some of the following variations in your students' graphs, some of which are apparent in the actual student graphs shown below and on pp. 25 and 27.

- Some students will write the exact times when people went in or out instead of using times at regular hour or half-hour intervals. Each method has advantages. Showing exact times allows the reader to tell a more accurate story, but it does not show the overall pattern of population change over time. Students will deal with this issue when they work out ways to combine information from their graphs in Sessions 3–4.

- Some students will supplement their graph by using keys, adding other illustrations, and writing explanations that they think the reader will need.

- Some students may use a symbol or a block on a bar graph for zero instead of leaving a blank. This is somewhat impractical, because a different symbol is needed to show one person at home. However, in the bar graph below where this has been done, the student's careful key does explain her system.

- Sometimes students will identify specific people instead of numbers of people. This approach may result in a chart. The graph below with people's names is structured as a chart, but it also functions very well as a graph. You can tell how many people are home at any given time, when people go in or out, and how long any one individual is at home.

Although nontraditional, all of these graphs and charts are "successful" according to the criterion for this activity: that someone, without additional information, can read them and know the number of people at home at given times during a day.

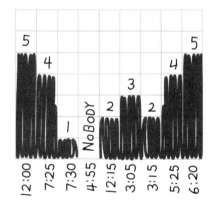

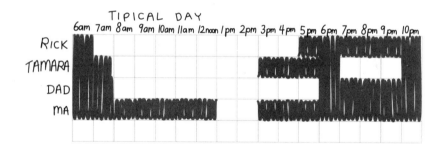

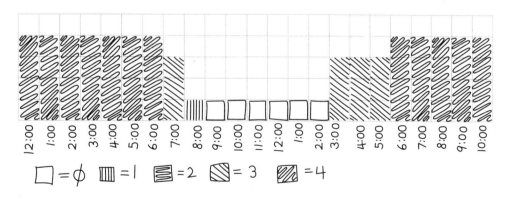

Graphing Home Population Over Time ■ **29**

Sessions 3 and 4

Making Group Graphs

Materials

- Plain paper and graph paper for each student
- Chart or poster paper (1 sheet per group)
- Colored pencils or pens
- Scissors, tape (1 per group)

What Happens

Students work in groups of four, each group making one graph that shows the changes of population in their four homes combined. Each group makes a poster to display the individual and group graphs they have made. Their work focuses on:

- developing a scale that includes all the data
- deciding how to group data
- establishing conventions for consistency

Note: Students continue to care for and measure their plants and record heights daily on Student Sheet 2, Plant Height Chart.

 Ten-Minute Math: Quick Images The Quick Images activity provides valuable practice in spatial concepts while students are working on this unit of data study. Two or three times over the next few days, outside of math hour, use the overhead to present patterns cut from the Quick Images Dot Pattern transparency, or patterns you have made yourself. To make your own, simply draw dots in several same-size groups, arranged in some pattern. For example:

Flash the dot pattern for 3 seconds and then cover it. Ask:

Can you draw the dot pattern you saw?

Give students time to draw what they remember, then flash it for another 3 seconds and let students revise their drawings. Ask:

Can you figure our how many dots you saw?

When most have finished drawing, show the pattern again and leave it visible for further revision and for checking the number of dots shown.

For full directions and variations, including Quick Image Geometric Designs, see pp. 82–83.

30 ■ *Investigation 1: Graphing Population Changes*

Activity

Combining Data on One Graph

Students get together in groups of four with the graphs they have made. Describe the task and allow time for them to discuss it.

Each group is going to work together to make one single bar graph that shows the movement of all the people, in and out of all four of your homes, as if you were all one big family.

Look over the data on your individual graphs and decide what you could do to combine it on one graph. On your group graph, you can identify different families by using different colors or symbols, but be sure your graph shows the total of people at home at any given time.

In their small groups, students decide how best to make a graph to show the changing population of the four households. Some groups may plan the graph first and then add data, while others may put all of their data together before they plan the graph.

While students are working, listen for problems that they raise. One likely dilemma is whether to write the times on the graph in regular intervals, say by the hour. If students do this, they will lose some detail. For example, people who leave home at 7:00, 7:25, and 7:45 will all be listed as leaving between 7:00 and 8:00. However, a regular scale is necessary for making a curve that can be easily interpreted. Some students may think of using fractions of hours, thereby preserving some detail.

You may need to help students decide how to combine their data and design their group graph. Talk with groups one at a time or bring the class together for a discussion of common issues as they begin to plan. They will need to establish some conventions in order to fit in all the data consistently. See the **Teacher Note,** Problems to Solve When Combining Data (p. 34).

Once the group has agreed on how to organize their data, they begin to make their graphs. They can tape two or more pieces of graph paper together if they need more space. If there are major disagreements, a group might make two alternative graphs displaying the same information. These group graphs will be displayed with students' individual graphs later in these two sessions.

Making Group Graphs ■ **31**

Activity

Assessment

Revising Individual Graphs

Students finish revising their individual graphs. These need not match the format of the combined graphs they are making in small groups; each student may use whatever plan seems to work best.

Assess the individual graphs for clarity and for the process students used to revise their work in response to their classmates' comments. Look for ways that students manage to combine the richness of their information with some consistency of format that makes the graph more readable.

In your assessment, do not focus on particular characteristics of the graph, but on how well it tells the overall story. For example, the graph does not need to have regular time intervals, but the reader should be able to tell at what time of day the graph begins, and should be able to follow in order the sequence of comings and goings and the total number of people at home throughout the day.

Activity

Assembling Posters

Each group of four now makes a poster displaying their group graph and their four individual graphs, with written comments under each graph. As a group, they plan some things to write about their group graph. For example, they might want to tell when the most people are home: "We have the most people at home between midnight and 6 A.M." Or, they might tell what the shape is and why: "Our graph goes up at both ends when lots of people are home and down in the middle when only six people are home."

Then all students write short paragraphs, comparing the way the group graph and their own graph display the same or similar information: "Four of the six people at home in the middle of the day are at my house when my mother takes care of my baby sister and two other children."

❖ **Tip for the Linguistically Diverse Classroom** Students who are not proficient in English can help illustrate the information on their group's poster with simple drawings. Instead of writing a short paragraph comparing their own graph to the group graph, students circle or mark in some way information on the two graphs that is similar, and list times of day when their graph is *not* similar to the group graph. Alternatively, students might write the paragraph in their native language.

32 ■ *Investigation 1: Graphing Population Changes*

Activity

Looking at the Overall Shape

You might pose the following questions to groups as they finish their posters, or to the whole class in a final discussion:

What is the overall shape of your group graph? Try sketching it without looking.

What would your group's graph look like if you made it continue for more hours? For more days?

Imagine a day for which your group graph would *not* be true. What sorts of days or events would cause a different pattern of ins and outs? (for example, weekends, holidays, days when school is dismissed early)

What do you think would be the shape of a graph of the population of all our households on a Saturday?

Collect the finished posters and display them where students can read them. If you do the Excursion (Sessions 5–6), some students will be telling stories from these posters. You will also want to refer back to these graphs during Investigation 3, Session 1.

Sessions 3 and 4 Follow-Up

Remind students to continue to bring in clippings that show changes over time to add to the class collection. Encourage them to look especially for graphs. They should discuss the clippings they collect with someone at home to be sure they understand them.

Teacher Note: Problems to Solve When Combining Data

Working in a group to combine data moves students to agree on which conventions to share. The group conversation becomes a negotiation of what to preserve and what to change from each individual graph.

Grouping data, if it is done effectively, makes overall trends stand out. For example, graphs of the combined population of several households will show a shape that is high at night when many people are home and low during the day when few people are home.

Students who are just learning about graphing sometimes combine their data in ways that fail to emphasize the overall trends. You might watch for problems like these:

- One group made a bar graph in which each student used a different color to put data on the graph. Each student colored in just the height for his or her own household, starting at the bottom of the graph. One student complained that his data had been buried: "B.J. has six people in his family and I only have four, so he just colored over my people at home and went higher."

- Instead of combining data for one "typical" day, a group of five students made a graph over the two different days for which the group members had collected data. Thus, two students showed their data on Tuesday and three showed their data on Wednesday, so there was no overall picture for one typical day.

- In another group, one student had begun his graph in the afternoon after school and continued it until the next afternoon. That group made a graph showing a full day from midnight to midnight and then continuing to the next afternoon, showing three students' data for the first day, four for overnight, and one for the next day.

Remind students that all of their graphs represent a *typical school day,* so the group graph can show only one day, from midnight to midnight. The graph should show the people in all of the households added together. One student explained it like this:

Alex: It doesn't matter about what day you did because you all did the same times, so you just write down for the day you covered. As long as you did 24 hours, it's the same thing.

You mean whether it's Tuesday or Wednesday, everyone has a 4 A.M. and a 4 P.M., so you just use the information from everyone in a graph of one day?

Alex: Yeah. We're just going to start at midnight and end at midnight.

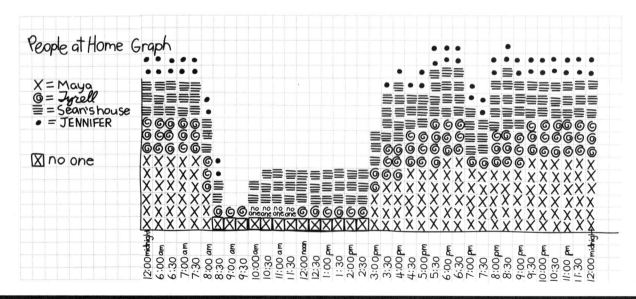

Sessions 5 and 6 (Excursion)

Ins and Outs Number Problems

What Happens

Students solve problems that involve finding a missing number in someone's report of the ins and outs from his or her home. Then students write fantasy story problems about people and other creatures coming and going, and trade problems to solve them. Since there is a lot of writing, rewriting, and reading involved, you might do this activity in writing time. Their work focuses on:

- understanding how changes and total are related
- developing strategies for solving missing information problems when the information is missing from the beginning, middle, or end
- writing missing information problems

Note: Students continue to care for and measure their plants and record heights daily on Student Sheet 2, Plant Height Chart.

Materials

- Students' home population graphs
- Cubes or chips (10 per student)

Activity

Ins and Outs Problems

Put out some small cubes or chips for students to use to model the ins and outs of story problems based on their home population graphs. Suggest that everyone have pencil and paper available, too. Begin by presenting a few problems yourself. Use real information from your home, or make up a story.

We're going to do some problems about people going in and out of their home. I'm going to tell you how many people are at home in the morning, and then how many go out and how many come in.

❖ **Tip for the Linguistically Diverse Classroom** To make the stories comprehensible, use simple drawings and arrows as you present the problems. You might begin by pantomiming to explain two simple pictures that you will use consistently to mean in and out. For example:

Here's the problem. When I wake up in the morning, there are five people at home, then one goes out, then three go out, then one comes back. How many people are in the house then?

Before we hear your answers, can anyone repeat the story?

Write down the starting number and the list of ins and outs, (for this example, 5 – 1 – 3 + 1) and ask if others agree. Then ask the students for their answers.

If the answers differ, choose one common answer and ask a volunteer to explain it. Then ask if any students have changed their minds about their answers, and what led them to change their minds. Encourage students to share all their different ways of solving the problem so they can get ideas from one another. The **Teacher Note,** Strategies for Ins and Outs Problems (p. 41), discusses some typical fourth grade strategies.

When everyone seems to understand how to do these problems, invite a student to stand near his or her graph and present a similar problem, telling how many people were home at the beginning, and then starting on the series of comings and goings. Stop the student after several changes and ask the class how many are in the house at that time. Then the student at the graph can confirm how many people were home. If students need more practice, do one or two more rounds.

Activity

Solving Backward Problems

When students are ready for a greater challenge, try some inverse problems.

This time, we will do a backward problem. I will tell you how many people go in and out and how many are home at the end. Then you will figure out how many were home at the *beginning*.

Model one problem for the class to do together. Start with a problem with two changes—one in and one out. For example:

Three people go out and one person returns home. Then there are four people at home. How many were home at the beginning?

As you say the problem, write it on the board:

$$? - 3 + 1 = 4$$

Collect all suggested answers and write them on the board.

We seem to have some different answers. It's fine to revise your opinions and change your answer at any time. I'm interested to know how you figured out your answer, and if someone else's explanation helps you to see the problem in a new way. Check your work by starting with the answer you got for the number at the beginning. Do each step, and see if the final number is correct.

If very few students are able to do this problem, try another problem with only one change. Again collect answers and ask students to explain their reasoning. Choose a student to give the next problem.

Pick a time on your graph to start. Mark the place, but don't tell how many people are at home then. Just tell how many went in and out.

Write a question mark on the board, and then the numbers of people who go in and out as the student reports them. After two or three changes, stop the student and ask how many are home at that point. Write that number on the board. Ask the class how many people they think were home at the beginning of the story.

Again, collect all the answers and write them on the board. Ask a different volunteer to check each answer out loud, starting with the number guessed and going through all the changes to check it.

The **Teacher Note,** Strategies for Ins and Outs Problems (p. 41), discusses some of the approaches to backward problems that fourth graders tend to use—some that work, and some that don't. It is much more difficult to find the beginning number than the end number. Students will need time and practice to work this out. Don't worry if many students can't do these problems after this brief introduction. Return to these problems from time to time after the unit is over.

Activity

Writing Story Problems

Ask students to write a fantasy story about what might happen to the population at their house during the day.

Start by telling the number of people and, if you like, animals, who are home just before you leave for school. Finish by telling the number of people (and animals) home just after you come home from school. The beginning and ending numbers can be the same or different. In between, tell who might visit your home while you are away. You may include fantasy people or creatures if you like. Be clear about the numbers that come and the numbers that go. If you want, you can include the times they go in and out.

Then take out or cover one piece of information to make your story into a math problem. Take out either the number at home at the beginning or the number at home at the end, or one of the numbers of creatures that went in or out while you were away.

Read the following problems as examples to get them started, or make up one or two problems of your own that will appeal to your class. Read each problem aloud twice—once for the students to get the overall story, and once for them to write down the numbers.

❖ **Tip for the Linguistically Diverse Classroom** Draw simple sketches on the board as you tell each story to aid comprehension.

Story problem 1 When my mom took me to the bus stop and she went to work, there were only our dogs in our house. But all the dogs in the neighborhood came through the dog door to visit—Fox, Shep, Tee, and Big Dog. That's four visitors. Then Tee ate all our dogs' food and left. Then Shep's owner, Mrs. Johnson, called, and Shep left and Fox left too. Then Fox came back. No other dogs came or went. When I got home, there were five dogs in the house. How many of them were ours?

38 ■ *Investigation 1: Graphing Population Changes*

Story problem 2   As far as we knew, the apartment was empty when we left for school. But then the flies came in through an open window. Eight hundred flies came into the apartment while we were away. My brother came home. He slapped at the flies and shooed them out and closed the window. He killed 13 flies, and he shooed 757 out of the apartment. *[Write these numbers on the board.]* When I came home, my brother was taking a nap. How many live flies were in the apartment? *[Or ask: How many live creatures were in the apartment?—in which case the storyteller and the brother should be included.]*

Each student writes one such problem. Students may use large numbers, but caution them to keep their stories manageable, using only a few changes, if they are going to ask people to guess the beginning number. After writing their stories, students remove one number and write a question asking for that number.

❖ **Tip for the Linguistically Diverse Classroom**   Students who are not proficient in English can create their stories in a storyboard format. They might include, where appropriate, the drawings you have used to indicate going in and going out.

> Hi, I live whith my Mother, father, older brother, older sister, great grandfather and five pet dragons in the small castle on the hill.
>
> One day when I was in a tower praktising wizardtree, my father, sister and grandfather went on a quest. They took 3 dragons. while they were away a group of 14 weery travlers stayed at our castle.
>
> How many people and Dragons are in the castle?
>
> 6 + 5 − 3 − 3 + 14 = 19

Ins and Outs Number Problems ■ 39

Activity

Solving Our Story Problems

In groups of four, students read their problems aloud for the rest of the group to solve. They must start by saying which number is missing: the beginning number, the end number, or one of the ins or outs. Suggest that students read their story once so others can get the idea, and then read it again slowly so others can write down the numbers and mark a place for the missing number. Students may use counters to keep track of the creatures.

Those students who are figuring out the missing number each find the answer, and then talk through their strategies until they all agree. Then the author of the story problem tells the answer.

❖ **Tip for the Linguistically Diverse Classroom** You or English-proficient students can read to the class the storyboards some students have created. Authors of the storyboards show the reader which number will be missing.

Sessions 5 and 6 Follow-Up

Story Problem Books Students might enjoy writing more story problems and combining their problems into a book. Use writing time for students to prepare final drafts for this publication.

Strategies for Ins and Outs Problems — Teacher Note

Keeping Track of Ins and Outs

Among fourth graders, we found three strategies that work for keeping track of the ins and outs when the missing information is the final number.

- Students might start with the initial number and add or subtract as each change is made, always keeping up with the current number at home.

- Students might make two lists of tallies, one for people going in and one for people going out. (Tallies for the number of people at home at the beginning go on the In list.) At the end, they subtract the total of the Out tallies from the total of the In tallies.

- Students might write down the initial number and then all the changes, in order, using plus and minus signs, and do the operations at the end:

  $5 + 2 + 1 - 4 - 2$

Do not try to influence the students to change their strategies. They will learn from each other and refine their strategies when the task demands it.

Solving Backward Problems

To solve backward problems, in which they must find the initial number, many students believe that they can just do the problem backward. *This does not work.* Take, for example, this problem:

> Four came in and two went out and then there were three. How many at the beginning?
>
> $(? + 4 - 2 = 3)$

Students who do this problem backward start with the end number and read from right to left:

> Three minus two plus four is five, so there were five at the beginning. $(3 - 2 + 4 = 5)$

If they check this answer, they will see that $5 + 4 - 2$ leaves 7, not 3. Students will find that the following strategies give better results:

- Students often adopt a workable trial-and-adjust method. They guess an initial number, then add and subtract through all the changes, and check to see if the end numbers match. If the numbers do not match, some students add to or subtract from the initial number according to the error and compute again to check. Others try guessing different initial numbers more randomly.

- Some students imagine the whole story and find the initial number by starting with the ending number and using the opposites of the changes. Take, for example, the problem mentioned above:

  Four came in and two went out and then there were three. How many at the beginning?
  $(? + 4 - 2 = 3)$

  Using opposites, one student explained, "Before the two people went out, there were five at home [3 + 2], and before four came in, there was one [5 – 4]."

- Some students collect the data, finding the total In and total Out, before adding the opposites. For example, consider this problem:

  $? +1 -3 +1 -1 +2 -1 = 6$ at home

  One student reasoned, "There were 5 out, 4 in. They started with 7 in, because 7 take away 5 is 2, plus 4 is 6."

  Or, as another student explained, "They started with 7, because 6 plus 5 [opposite of 5 out] minus 4 [opposite of 4 in] is 7."

INVESTIGATION 2

Ways to Show Change Over Time

What Happens

Sessions 1 and 2: Showing Change Over Time
Students examine a variety of items from newspapers, magazines, and other sources that relate to changing situations and events. They sort these examples into two piles, those that show actual changes and those that do not. Students invent their own ways to show changes in everyday events in their lives.

Mathematical Emphasis

- Examining real situations and events that show change
- Interpreting representations that show change
- Making representations that show change
- Distinguishing between representations of something that can change and representations that show changes

INVESTIGATION 2

What to Plan Ahead of Time

Materials

For ongoing measuring:
- Plants in pots (all sessions)
- Centimeter rulers: 1 per pair (all sessions)
- Students' Plant Height Charts for daily recording

For the regular investigation:
- Plain paper: 2 pieces per student plus extras
- Overhead projector (for Ten-Minute Math)

Other Preparation

- Collect newspaper weather maps from three consecutive days, choosing maps that show a weather front moving across the country. Combine the maps on a transparency for class viewing or on a page to photocopy, 1 per 2–3 students.

- Gather the clippings that you and students have brought in that show how things change. Ideally, you should have 6–10 clippings to distribute to each small group of students. Some of these may not actually show change, but only the *result* of change; this is fine, as students will be sorting them in Session 1.

  Select an area where you can display the clippings after Session 1. Put up two headings: *These do not show change over time* and *These show change over time*. You might display a few clippings under each heading as examples. **Note:** To show change, the clippings must illustrate two or more different situations. See p. 45 for further explanation.

- For Ten-Minute Math, have available the cut-apart Quick Image transparencies you prepared during Investigation 1.

Investigation 2: Ways to Show Change Over Time

Sessions 1 and 2

Showing Change Over Time

Materials

- Clippings that show change
- Weather maps for three consecutive days
- Plain paper

What Happens

Students examine a variety of items from newspapers, magazines, and other sources that relate to changing situations and events. They sort these examples into two piles, those that show actual changes and those that do not. Students invent their own ways to show changes in everyday events in their lives. Their work focuses on:

- interpreting representations that show change
- distinguishing between representations of something that can change and representations that show change over time
- making representations that show change

Note: Students continue to care for and measure their plants and record heights daily on Student Sheet 2, Plant Height Chart.

Ten-Minute Math: Quick Images Two or three times in the next few days, continue to use one or two images from the Quick Image Dot Patterns transparency or some that you or your students draw. Do this activity outside of math time.

For full directions and variations, see pp. 82–83.

44 ■ *Investigation 2: Ways to Show Change Over Time*

Activity

Looking at Ways to Show Change

Change on Weather Maps Show the first of the three weather maps you have collected. Ask students what they can learn by looking at one map (such as what the temperature is in different places, where it is raining, where it is sunny). Show the second map and ask what else students can learn by comparing maps from consecutive days (such as how the temperature has changed in a particular place, where the rainstorm has moved).

Point out that although one weather map has a lot of information on it, it does not actually show change over time. Here you might discuss how the weather forecasters on TV show change with one map. Do they really use only one map? How do they show changes? Students then look for changes revealed by comparing all three weather maps.

Sorting Clippings Distribute the clippings about change, giving 6–10 to each group of four students (two pairs). Try to give each group a variety of formats—graphs, cartoons, charts, and so on.

These clippings you collected show things that change, but only some of them show the actual change. Let's look at them to see what they're really showing.

Students sort the clippings into two piles, those that show change, and those that do not. As needed, provide some examples to help students see the difference.

These do not show change over time	*These show change over time*
■ a single weather map	■ weather maps from consecutive days
■ an ad that reads, "prices slashed"	■ an ad with both regular and sale prices
■ a new high jump record	■ high jump records over several years
■ a photo of spring flowers	■ a photo of winter snow next to a photo of spring flowers
■ a wedding photo	■ a wedding photo next to a photo of the couple before the wedding

The issue is to distinguish between what we know by inference, because we recognize the results of a change, and what the illustration actually shows.

Showing Change Over Time ■ **45**

As students do the sorting activity and discuss their choices, they will establish and clarify for themselves a definition of change and will note that to show change, there must be a "before" and an "after," or an "earlier" and a "later."

When a group has agreed about the placement of their clippings, they post the clippings on the prepared display area in the two categories. Students can post stick-on notes with their initials on clippings they think have been placed in the wrong group. Later, have a brief discussion about the clippings in question.

Activity

Showing Change Our Own Ways

Students find their own way to tell how something they know and care about changes—such as their home, their family, or themselves. They might make a graph, or put together a series of pictures, cartoons, or three-dimensional constructions.

Students can brainstorm some ideas of changes they might depict. Write their ideas on the board. You might start by listing *ways* things change—in speed, size, age, color, amount, and so forth. Then students could focus on things that change in those ways. For example, here are some ideas students have worked with:

- their own height
- a sister's or brother's changing hairstyle
- the life story of a favorite piece of clothing or a baby blanket
- the different noises in their home over an evening
- their favorite foods or books or sports at different ages up to now
- the changing appearance of a baby brother or sister, or of a pet
- changes in a favorite tree over a year
- improving records in a favorite sport over several years

Students can get ideas for how to show change from the class display of clippings, or they can make up new ways. This is a low-key activity, with students talking quietly as they work.

Find a place to display the finished products.

Sessions 1 and 2 Follow-Up

Students can finish the representation they are working on or make another in a different format.

🏠 **Homework**

Displaying Ways of Showing Change Encourage students to continue to bring in examples of interesting ways of showing change over time to add to the class display. They should particularly look for line graphs, as they will be making line graphs to show the growth of their plants in the next investigation. You might even suggest that students bring in *any* graphs they find, whether or not they show change. Then you could add a third category on the display, for graphs that are not about change at all.

◩ **Extension**

Showing Change Over Time ■ **47**

INVESTIGATION 3

Telling Stories from Line Graphs

What Happens

Session 1: Making Graph Sketches Students compare the shapes of two graphs—the changing population of their homes and of their school over 24 hours. They then find a way to show changing speeds as described in a wheelchair racer's story of the Boston Marathon.

Session 2: Graphing and Predicting Plant Growth Students start a line graph of their plants' changing heights. They predict and sketch what a graph of their plants' continued growth might look like. Then they graph further actual heights and compare the graphs of predicted and actual growth. They see how the shape of their graph indicates fast and slow growth.

Session 3: Using Line Graphs to Compare Growth Students from two different groups put their two graphs of plant growth on the same set of coordinates, solving problems of scale in ways that make sense to them. They compare their graphs to see where each plant has grown fastest and slowest.

Session 4: Graphs, Stories, and Number Sequences Students match graphs with sets of plant heights and descriptions of plant growth.

Session 5: Interpreting Graphs Students look at a variety of graphical shapes that represent the growth patterns of different plants and describe in words how each plant grew.

Sessions 6 and 7: Mystery Graphs Students make line graphs that show how the population changes throughout a day in different (unnamed) places or activities. The class tries to match the mystery graphs to a list of places and activities. Students revise their graphs to clarify anything that was confusing, and make a display to post outside the classroom.

Mathematical Emphasis

- Attending to the shape of graphs
- Using curves to communicate information
- Developing an understanding of the difference between continuous and discrete changes
- Describing growth with a line graph
- Comparing graphs of slow growth and fast growth
- Developing a scale that includes all the data
- Making, interpreting, and comparing line graphs
- Integrating quantitative, qualitative, and graphical descriptions of the same data
- Telling a story from a curve
- Making and interpreting different graphical shapes

INVESTIGATION 3

What to Plan Ahead of Time

Materials

For ongoing measuring:
- Plants in pots (Sessions 1–4)
- Centimeter rulers: 1 per pair (Sessions 1–4)
- Students' Plant Height Charts for daily recording
- Sticks and string, to support the students' plants as they grow taller

For the regular investigation:
- Plain paper and graph paper for each student
- Glue or tape (Sessions 1, 4)
- Colored pencils or pens (Sessions 2–3)
- Scissors: 1 per student (Session 4)
- Large paper (11" by 17" or larger): 1–2 per student
- Light and dark marking pens: 1 each per group (Sessions 6–7)
- Overhead projector
- Calculators (Ten-Minute Math)

Other Preparation

- Duplicate student sheets and teaching resources (located at the end of this unit) as follows:

  For Session 1

  Student Sheet 4, The Boston Marathon: 1 per student

  Motion Graph (p. 103): 1 transparency

  For Session 2

  Student Sheet 5, Fast and Slow Growth: 1 per student, and 1 transparency

  Student Sheet 6, Our Growing Plants: 1 per student

  For Session 4

  Student Sheet 7, Matching Numbers, Stories, and Graphs: 1 per student

  For Session 5

  Student Sheet 8, Plant Graph Stories: 1 per student

  For Sessions 6 and 7

  Student Sheet 9, Mystery Places and Activities: 1 per student

- Cut up a copy of Student Sheet 9 so that each listed item is on one slip of paper. If you have a linguistically diverse classroom, add simple drawings or real items (a movie ticket stub, a pizza coupon, a library card) to help identify the items.

- Make an enlarged copy of the list on Student Sheet 9 to post. As needed, use the same visual aids suggested for the cut-apart slips.

- Make a transparency or photocopies of a typical child's growth curve (check with a pediatrician or in child development books). (Sessions 6–7, optional)

Investigation 3: Telling Stories from Line Graphs

Session 1

Making Graph Sketches

Materials

- Paper for each student
- Glue or tape
- Student Sheet 4 (1 per student)
- Transparency of Motion Graph
- Overhead projector

What Happens

Students compare the shapes of two graphs—the changing population of their homes and of their school over 24 hours. They then find a way to show changing speeds as described in a wheelchair racer's story of the Boston Marathon. Their work focuses on:

- attending to the shape of graphs
- using curves to communicate information

Note: Students continue to care for and measure their plants and record heights daily on Student Sheet 2, Plant Height Chart.

 Ten-Minute Math: Broken Calculator Three or four times during this investigation, use the Broken Calculator activity. This can also be assigned for homework. Pose a problem such as this:

Imagine that the plus and minus keys on your calculator are broken. Devise a problem that will result in the answer 30 on your calculator without using + or −.

After students solve the problem, list some of their solutions on the board. Ask students to choose one solution and extend it to form a group of similar solutions or a series that follows a pattern.

For example, to form 30 without using + or −:

1×30	$30 \div 1$	$30 \times 1 \div 1$
2×15	$60 \div 2$	$30 \times 2 \div 2$
3×10	$90 \div 3$	$30 \times 3 \div 3$
4×7.5	$120 \div 4$	$30 \times 4 \div 4$
5×6	$150 \div 5$	$30 \times 5 \div 5$

For full directions and variations on the Broken Calculator activity, see pp. 84–85.

50 ■ *Investigation 3: Telling Stories from Line Graphs*

To introduce line graphs that show change over time, begin by briefly reviewing some graphs that are familiar to students. Remind students about the graphs they made in Investigation 1, showing how the population in their homes changed over 24 hours. Call attention to one of their group graphs, showing the population of four homes together. Ask them to describe the overall shape of the graph. (It is approximately U-shaped.)

Sketch different shapes of graphs on the board and ask:

Did it look like one of these?

Is it higher in the morning or in the afternoon or at night? Is it almost the same all day? Does it have two humps or only one?

Take student suggestions until everyone agrees on the shape.

Use a single line to sketch the graph of the population of your homes. Consult with your partner.

When students have finished, sketch the shape on the board and label it.

Now ask your students to all think about how a graph of the population of the school over 24 hours (on a school day) would look. (It would look approximately like an upside-down version of the home graph.) Give them a minute to sketch it and talk with their partners about it. Then take suggestions. When you get agreement, draw it on the board also.

Activity

Discussing Overall Shapes of Graphs

Ask the following questions about the home and school population graphs:

What does a line going high up mean on each of these graphs? (more people) **What does a line going low down mean?** (fewer people) **What does the very lowest part of the line mean?** (no people)

Activity

Describing a Marathon Graphically

Pass out two pieces of paper and tape to each student as you introduce this activity. Students tape the pieces together along the short sides and place the paper in front of them with the longer dimension running from left to right. (Some students may need to extend their graphs to the right with a third sheet.)

I'm going to read a story written by someone who has been a wheelchair racer in the Boston Marathon since 1975. As I read, pay attention to how the racer's speed changes during the race.

I'll read the story two times. The second time, I'll read slowly, stopping from time to time while you draw a line graph on your long paper to show how the racer's speed is changing.

Alert the students to a common confusion that arises during this activity:

Some people get confused about what to do when the racer goes up or down a hill, and they make their line graphs look like a hill. Remember, you are drawing how the racer's *speed* goes faster and slower, not the way the race course goes up and down.

Encourage students to think about how they might do this.

In our graphs about population, "high up" meant more people and "low down" meant fewer people. What will high up mean in a graph about the racer's speed? (going fast) **What will low down mean?** (going slowly) **What will the lowest point mean?** (stopped)

You will hear the racer describe the hills he had to go up. Remember you are not graphing the hills, but how fast or slow the racer was going.

Graphing the Racer's Speed Read the story (p. 53) through once at normal speed. After reading the account once, ask students how they can show on a graph the difference between going at a moderate speed and going fast. They should all have the idea that going fast and going at a medium speed would both be shown with horizontal lines, with the line for going fast drawn up higher.

The Boston Marathon

Participating in the Boston Marathon—all 26.2 miles of it—is an incredible experience. You're with thousands of other people, going through all kinds of different towns and cities and college campuses. Wheelchair participants get a lot of attention at Boston. People know us and they start us right at the front.

In the beginning of the race, there's about 4 miles of downhill. Like most wheelchair participants, I **wheeled very fast** on that downhill part. It felt good. But at mile 4, I remembered to pace myself. I **slowed down a bit.** For the next 9 miles or so, I **wheeled nice and steady.**

But then a very exciting thing happened. At Wellesley College, there was a huge crowd of students lining the course, screaming and clapping like crazy. All that excitement made me **push faster and faster** through the mile-long part of Wellesley. But then I noticed I was tired—too tired for being just a little more than halfway through the race. I **stopped** for a few seconds to get some water and pour some over my head. (I was getting hot, too!) I knew the hardest part of the race was coming up.

After Wellesley, I **wheeled steadily** for a few more miles. But then, by the seventeenth mile, it started getting hard. From mile 17 to 21 or so, I could feel myself **gradually slowing down.** There's a bunch of hard hills, and I knew I just had to take it easy to make it over those hills. My arms were aching so much. But the funny thing was, even once I had made it over the hills, I **kept pushing slowly.** I think by 21 miles I was running out of steam. It was hard to keep picking up my arms. Between 21 and 25 miles, I kept pushing slowly.

By 25 miles, I knew I would make it and I **picked up the pace.** The crowd was tremendous in the last mile or so. They just wouldn't let you slow down. The final stretch of a quarter mile or so is downhill, and I actually **ran my fastest** for that stretch. My arms felt all beaten up and shaky at the finish—but I **wheeled around slowly** for a while afterwards. That helps you keep from getting so stiff the next day.

Students may be interested in knowing that the information in this story was provided by Bobby Hall of Cambridge, Massachusetts, who was the first recognized wheelchair participant in the Boston Marathon. Mr. Hall won the wheelchair division in 1975 and 1977, and has pioneered wheelchair racing both in Boston and across the world. Mr. Hall's marathon time in 1975 was 2:54:05, almost three hours.

Today, the best wheelchair racers are wheeling times that are twice that fast (less than 90 minutes!). The faster times are a result of more and better training, the involvement of more and more wheelchair athletes, and improvements in their equipment.

Then read the story more slowly, pausing between sections for students to graph what the story has described. The **Teacher Note**, Graph Sketches of Speed (p. 56), describes some of the problems students enounter.

❖ **Tip for the Linguistically Diverse Classroom** For the second reading, ask an English-proficient student to act out the part of the wheelchair racer, showing the changing speed through hand motions. Be sure to have the student demonstrate distinctly speeding up, slowing down, and moving at a steady fast or slow speed.

Activity

Interpreting the Graphs

Adding Words to the Graphs Pass out a copy of Student Sheet 4, The Boston Marathon, to each student. Read the instructions for Part 1. Ask your students to find the places on their graphs where the racer is doing what the underlined words or phrases describe, and to write the words or phrases in the appropriate places along their curves. Do the first one or two phrases together to be sure everyone understands the task (see the drawing in the **Teacher Note**, Graph Sketches of Speed, p. 56).

Encourage students to help each other as they work. They may adjust their lines as needed to fit the words better. When they are finished, they can look over each other's graphs to see if the labels seem to fit with the parts of the graph and if they understand the graph as a whole.

❖ **Tip for the Linguistically Diverse Classroom** Read the marathon description aloud for students who need this help. Stop after each underlined word or phrase. Ask English-proficient students to work with other students to find the appropriate places on their graphs. Model on the board what they are to write.

Ways of Describing the Speed and Changes in the Speed Read the instructions to Part 2 of Student Sheet 4. Make sure students understand the task: to sort the words and phrases into two lists—words that describe a specific speed, and words that describe a change in speed. Again, you may want to do one or two of these together, and encourage students to help each other. See the **Teacher Note**, Speed Versus Change in Speed (p. 57), for ways to clarify the distinction between steady speed, whether fast or slow, and change in speed.

❖ **Tip for the Linguistically Diverse Classroom** Again, you might have a student act out moving at different speeds in a wheelchair as you read the items on the list in Part 2 of Student Sheet 4.

Class Discussion When everyone has finished Student Sheet 4, ask your students how they showed the racer going at different speeds on their graphs. Invite them to come to the board to draw parts of the graph that show the racer's pace increasing or decreasing and to explain their thinking.

How do you show the racer wheeling pretty slowly? How does this compare to the way you showed him moving faster? What does "stop" look like on your graph?

Move on to words and phrases that describe a change in the racer's speed.

How did you show him pushing faster and faster? How did you show him slowing down? What does picking up the pace look like? How does changing pace look different from going at a steady pace?

"Wheeled nice and steady" and "wheeled steadily" seem to mean the same thing. How does your graph show this?

Did anyone's graph show a difference between stopping suddenly to get a drink and slowing down gradually because the racer is tired?

Activity

Another Graph of Motion

Display the transparency of the Motion Graph (p. 103) and tell students that it shows a race different from the one they have just graphed. Students can work in small groups and quietly discuss the questions on the transparency, deciding on the answers together. When they are ready, bring the class together to report and discuss their answers.

Notice that the racer goes fast for more than half the race, then stops, then goes more slowly for the rest of the race.

What might have happened to this racer to result in this line?

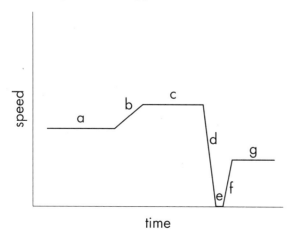

Making Graph Sketches

Teacher Note: Graph Sketches of Speed

We have found that when fourth grade students generate their own graphs for a story of motion, they use picture and color codes or different lines (e.g., wavy for fast and straighter for slow) to show different speeds. They arrange these in a straight line. The activity based on the wheelchair racer's story is designed to encourage students to start using a line graph that moves *up* and *down* while also moving *across* the page to show how something changes.

In this racing story, the wording suggests critical distinctions to be made on the graph. For example, horizontal lines that are at different heights can be used to distinguish between going slowly, going at a moderate pace, and going fast. Discourage students from using codes for different speeds. Suggest they try to show different speeds only with different heights of the line.

Due to the qualitative nature of the racer's description, your students' interpretations will differ, both in terms of the racer's speed (how high the graphs are at a given point) and in terms of the length of time the racer does something (how far across the graph the lines go). Keep in mind that the goal is for the students to make an overall representation of how the racer's speed changed over time.

One interpretation of the racer's speed might look like the line graph below.

Even with initial warning, some students may still be tempted to make their graphs match the description of the race course rather than the speed of the racer. That is, their graphs will go up when the racer is going up a hill, or down when the racer is zooming downhill, even though the opposite would be the best way to show changing speed. We call this phenomenon "graph as picture." Expect this kind of confusion. Discuss it when it comes up. This activity is a good opportunity for students to become aware of this potential pitfall.

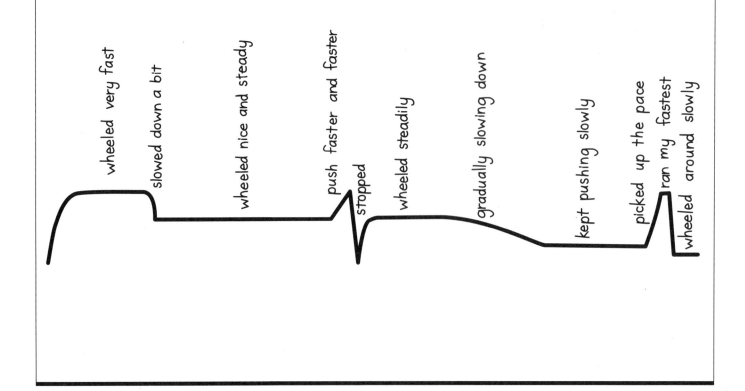

Speed Versus Change in Speed

> Teacher Note

One of the goals of the activity Describing a Marathon Graphically (p. 52) is for students to distinguish between what a graph of speed looks like when it describes a steady motion *(kept pushing slowly, wheeled very fast)* and what it looks like when it describes a change in motion *(push faster and faster, gradually slowing down)*. In general, when a graph of speed describes going at a certain pace, the line is horizontal; when it describes changing pace by speeding up or slowing down, the line is at a slant, going up or down. Students who make this sort of graph will show a faster steady pace as a higher horizontal line and a slower steady pace as a lower horizontal line. A few students will distinguish between slowing down quickly and slowing down gradually by drawing a steep straight line or a less steep line.

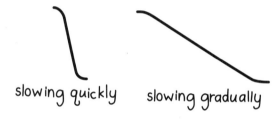

In the racer's story, students usually agree that these words and phrases describe a steady pace: *wheeled very fast, wheeled nice and steady, wheeled steadily, kept pushing slowly, ran my fastest, wheeled around slowly.*

These are the phrases that most students think describe a change in pace: *slowed down a bit, push faster and faster, gradually slowing down, picked up the pace.*

Students disagree about where to put the word *stopped*. They can make some good arguments on both sides.

Sarah: I put *stopped* in the change list because first he was moving and then he stopped.

Sarah is thinking that you must be doing something before you stop.

Karen: You might have been going fast before you went slowly, but going slowly is not a change in speed, so stopped is not a change in speed.

Nhat: If you say a car is stopped at the curb, it is not changing speed.

Sarah: But before he stopped, he slowed down. You can't just go fast, stop.

Think about how your graph shows that he slowed down before he stopped. Think about how it shows that he stopped for a moment. How would it look if he stopped for a longer time?

Making Graph Sketches ■ **57**

Session 2

Graphing and Predicting Plant Growth

Materials

- Student Sheet 5 (1 per student)
- Transparency of Student Sheet 5
- Student Sheet 6 (1 per student)
- Graph paper
- Colored pencils or pens
- Overhead projector

What Happens

Students start a line graph of their plants' changing heights. They predict and sketch what a graph of their plants' continued growth might look like. Then they graph further actual heights and compare the graphs of predicted and actual growth. They see how the shape of their graph indicates fast and slow growth. Their work focuses on:

- developing an understanding of the difference between continuous and discrete changes
- describing growth with a line graph
- comparing graphs of slow growth and fast growth

Note: Students continue to care for and measure their plants and record heights daily on Student Sheet 2, Plant Height Chart.

Activity

Graphing Our Plant Heights

Different Kinds of Changes in Data Ask your students to compare the kind of data they collected during Investigation 1 with the data about plants they are currently collecting. Point out that the changing populations of their homes can be expressed only in whole numbers (whole people), but the changing plant heights can be given in fractions (parts of a centimeter). The **Teacher Note,** Continuous Versus Discrete Changes (p. 63), discusses further the difference between these two types of data and the implications for graphing them.

What else can you think of that, like population, changes only by whole numbers?

List students' ideas on the board (for example, number of books in a library, number of cars in a parking lot, number of baseball cards collected).

And what other things change by fractional numbers, like the plant height data?

Again, list students' ideas on the board (for example, volume of water in a lake, temperature, hours of daylight, lengths of time, speed).

Starting the Graphs Throughout this investigation, unless otherwise specified, students who share a plant work together. However, each member of the team produces his or her own graphs and other written work.

So far, you have been keeping track of your plants' heights in a chart. Today you will make a *line graph* that shows how your plant has been growing.

Draw the axes of a graph on the board. Tell your students that all the days of the week will go along the horizontal axis, and point to it.

What would go along the vertical axis? [*Point to it.*] **What will "up high" mean on this graph? What will "down low" mean?**

Agree that plant heights will go along the vertical axis. See the **Teacher Note**, Students' Line Graphs (p. 64), for a discussion of some other conventions to discuss with students.

Distribute graph paper. Students draw axes on their graphs. They write height in centimeters along the vertical axis and days along the horizontal axis. At this point, students graph just the first five heights of their plants from their plant growth charts. They write the days at the bottom of the vertical lines on the graph. They put a point on each of these vertical lines to correspond with the height and day for each of the five measurements. They then join the points with line segments in colored pencil. Ask students to think about what they can do for days when they have no measurements—weekend days or days when they are absent or forget to measure.

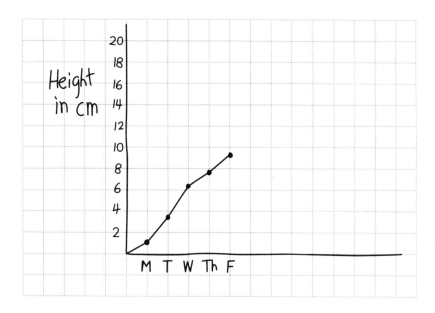

Activity

Predicting Future Growth

Put your Plant Height Chart aside and just look at your graph. Over the next week, you'll keep on measuring your plant daily and graphing its growth. What do you think your graph will look like after another week?

Working on the line graphs they have just made, students use a different pencil color to draw what they predict the graph of their plant's growth will look like when they add measurements for one more week. Pairs of students may decide this together. Then they explain their thinking individually on notebook paper.

❖ **Tip for the Linguistically Diverse Classroom** Instead of a written explanation, students might draw their plant in stages as they have predicted it will grow during the next week.

Some students may think their plant will continue to grow in the same way it has grown until now. Others may think the growth will speed up or slow down. The accuracy of their predictions is not important here. What is important is that students begin to think about how plants might grow and how to represent growth on a line graph.

my plant is growing good and I think it is going to grow sky high next week because of the warm spring sun.

Activity

Faster and Slower Growth

Distribute Student Sheet 5, Fast and Slow Growth. Note that the graph on the student sheet shows every day of the week to maintain the scale of moving one unit to the right for each day, but the weekend days have no measurements (no dots on the graph).

As a class, read through the questions about the graph. Allow about five minutes for students to discuss the questions with their partners and jot down answers. Then display the transparency of Student Sheet 5. Direct students' attention to the first weekend.

Some people think the plant grew fastest on this weekend. Other people think it grew fastest from Monday to Tuesday. What's your thinking about that?

Day	Height
Thursday	3 cm
Friday	4 cm
Monday	10 cm
Tuesday	13.5 cm
Wednesday	15 cm
Thursday	16 cm
Friday	16.5 cm
Monday	17.5 cm

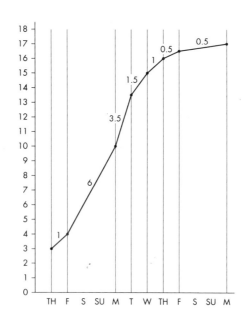

Discuss the rest of the questions with students.

Where does this graph show fastest growth? (between Monday and Tuesday) **How can you tell from the shape of the graph?** (It's steepest there.) **How can you tell from the numbers along the graph?** (Since the 6 centimeters is over three days, the plant grew only 2 centimeters per day. However, between Monday and Tuesday, the plant grew 3.5 centimeters.)

Where does the graph show slowest growth? (between Friday and Monday at the end) **How can you tell from the shape of the graph?** (It's flattest there.) **How can you tell from the numbers along the graph?** (The change is only half a centimeter for three days.)

Graphing and Predicting Plant Growth ▪ **61**

Continuing the Growth Graphs Students take out their own graphs showing the actual growth of their plants as well as their predictions. Using their original pencil color, they add to their graphs as many actual measurements as they have to date. Depending on how much actual data your students have at this point, they will be able to begin comparing their predictions with their plants' actual growth.

Activity

Teacher Checkpoint

Our Growing Plants

Distribute Student Sheet 6, Our Growing Plants. Students should still have their plant height graphs at hand. First they write the changes in growth, in centimeters, along their line graphs (as was done on the graph on Student Sheet 5). They then answer the questions about their own graphs.

❖ **Tip for the Linguistically Diverse Classroom** Have students circle the part of their graph and the measurements on their chart that indicate *fast* growth, and box the part of their graph and the measurements that indicate *slow* growth.

As students are working, check their explanations to see if they understand how both the slope of their graphs and the numbers showing the changes reflect how fast their plants are growing.

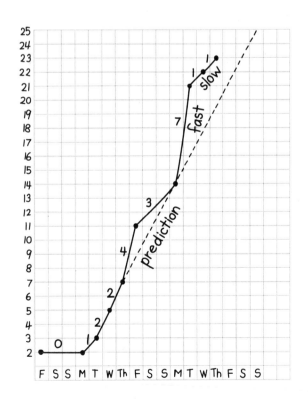
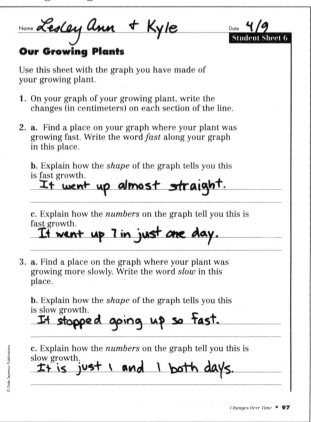

62 ■ *Investigation 3: Telling Stories from Line Graphs*

Session 2 Follow-Up

Assign a Broken Calculator problem like the ones students have been doing for Ten-Minute Math. Pick a number that has many factors, such as 48 or 72. Ask students to make a list of similar solutions to arrive at that number without using the plus or minus key. For example:

72	**48**
2×36	2×24
4×18	$2 \times 2 \times 12$
6×12	$2 \times 2 \times 2 \times 6$
8×9	$2 \times 2 \times 2 \times 2 \times 3$
10×7.2	$2 \times 2 \times 2 \times 2 \times 2 \times 1.5$

Continuous Versus Discrete Changes — Teacher Note

During the first two investigations, students worked with changes that jumped from one whole number to another without passing through all the numbers in between. The number of people at home was always expressed as a whole number—2 or 4, for example, never 2.3 or 4.1. A bar graph captures this *discrete* nature of changing populations.

Changes in plant size, on the other hand, involve all the numbers between two different measurements. A plant that measures 6 centimeters today and 7 tomorrow was every possible size between 6 and 7 centimeters. A line graph captures the *continuous* nature of such data. During this investigation, we want students to begin thinking about the continuous nature of plant growth.

Your students may think of a case when it would be appropriate to express the number of people as a fraction or decimal—when a person is part way into a room, for example. It is important to recognize the validity of this case if it comes up, and to let students know that this is an example of how the distinction between continuous and discrete data is not always clear.

Give students time to consider how keeping track of data and graphing them is different for the plants and for the home population. Here is a bit of the discussion of the topic from one fourth grade class:

What is the difference between keeping track of the people at home and the heights of your plants?

Tyrone: When you do the population change, it is like counting How many, and when you do the plant, it is How tall? How big?

What is the difference between the numbers?

Jesse: We're using *times* for people.

Lesley Ann: But they're both times—days for plants, and hours for home.

Kim: If they are plants, the numbers go up, not down. If it is population, they go both up and down.

Shiro: For plants, you don't know how big it is going to be, but with people, you know how many is usually the most.

Teacher Note: Students' Line Graphs

When students began graphing their population data in Investigation 1, they tended to include the exact information they had. They display this same tendency when graphing plant height data. Along the x-axis, for example, they typically indicate only the days on which they measured their plants, leaving out the weekend days. The chart below shows the heights of a plant that was measured on Friday and then again on Monday.

Day	Height
T	2 cm
W	2.5 cm
Th	3 cm
F	4 cm
M	7 cm
T	8 cm
W	8.5 cm

The graphs below show two different ways to graph these data. In the first graph, where the weekend days are not included, the plant appears to have had a growth spurt over the weekend. In the second graph, where the days of the weekend are included, the plant appears to have continued to grow at a steady rate. In order to avoid creating a misleading impression about how their plants are growing, students need to put *all* the days of the week along the x-axis.

Students also tend to start numbering along the y-axis with the exact measurements of their plant, rather than establishing a scale with even increments. For example, if their first measurements are 3.5 cm and 5 cm, they will make 3.5 and 5 the first numbers along the y-axis. Allow them latitude at this time. Later in the investigation, they will look more closely at the consequences of this practice.

Whenever you introduce a conventional representation system, encourage students to discuss how it differs from the systems they have invented. These discussions should help students become aware of the benefits of conventional approaches, such as Cartesian graphs—graphing points (x, y) on the coordinate plane—that people have constructed through history, without discounting the benefits of their own invented approaches, such as putting down all the times when people came and went on their population graphs.

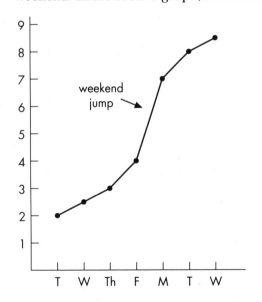
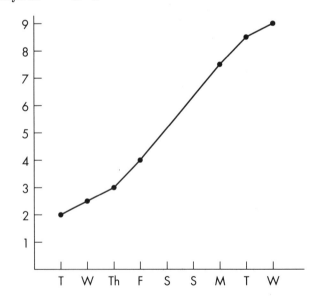

64 ■ *Investigation 3: Telling Stories from Line Graphs*

Session 3

Using Line Graphs to Compare Growth

What Happens

Students from two different groups put their two graphs of plant growth on the same set of coordinates, solving problems of scale in ways that make sense to them. They compare their graphs to see where each plant has grown fastest and slowest. Their work focuses on:

- developing a scale that includes all the data
- making, interpreting, and comparing line graphs

Note: Students continue to care for and measure their plants and record heights daily on Student Sheet 2, Plant Height Chart.

Materials

- Graph paper
- Colored pencils or pens

 Ten-Minute Math: Broken Calculator Continue to give students Broken Calculator problems. You might ask them to make given numbers without using the digits in those numbers; for example, to make 99 without using a 9, or make 200 without using a 2 or 0.

After students solve the problem, list some of their solutions on the board. Ask them to choose one solution and extend it into a series that follows a pattern. For example:

99	200	200
100 – 1	199 + 1	1 × 5 + 195
101 – 2	197 + 3	3 × 5 + 185
102 – 3	196 + 4	5 × 5 + 175
103 – 4	195 + 5	7 × 5 + 165

For full directions and variations, see pp. 84–85.

Activity

Combining Graphs of Different Plants

Students switch partners so that each new pair has two different graphs.

Today you will compare the ways two of your plants have been growing by putting both on the same grid.

You will need to decide together what to put along each axis of the new grid. Once you have decided this, graph both plants on the same grid using a different color for each plant. Both members of a pair work on the same piece of paper, so each pair makes only one set of graphs.

Together students work out how they will represent their two graphs on the same set of coordinates. Once a pair and their regular partners have finished, the foursome get together and compare their ways of graphing the two plants and the ways the plants grew. They consider the following questions:

What numbers did you put along the side for heights of the plants? What is similar about the growth of the two plants? What is different about their growth?

Students take a few minutes to look over each other's graphs and discuss these questions.

Once students have talked in their small groups, ask them to discuss as a class the process of putting their graphs together on the same grid.

To plan these graphs, what things did you have to agree on? What was hard to decide?

Session 3 Follow-Up

Assign another Broken Calculator problem. Students show a pattern of ways to get the answer.

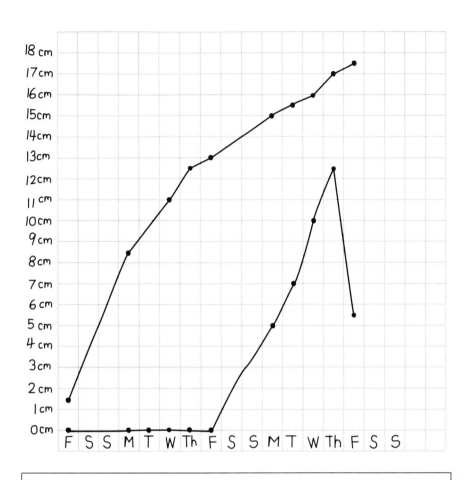

1. We put the cm. going up and the days going across. Our plants were shone in two differant colors.

2. They both went sudly up when they started to grow.

3. David's plant stayed in the growned for 9 days but mine grew right away and got much taller then his.

4. We had trouble tring to graph David's plant because it broke but we firged out how to do it.

Session 4

Graphs, Stories, and Number Sequences

Materials

- Student Sheet 7 (1 per student)
- Scissors
- Glue or tape
- Paper for mounting (1 per student plus extras)
- Poster paper (1 per student, optional)

What Happens

Students match graphs with sets of plant heights and descriptions of plant growth. Their work focuses on:

- integrating quantitative, qualitative, and graphical descriptions of the same data
- telling a story from a curve

Note: Students continue to care for and measure their plants and record heights daily on Student Sheet 2, Plant Height Chart.

 Ten-Minute Math: Broken Calculator Continue to pose Broken Calculator problems over the next few days, perhaps asking students to make a small number such as 2 with only multiplication and addition, or a decimal such as 0.25 with only subtraction and division. See decimal variation, p. 84. Also continue to assign some of these problems for homework.

68 ■ *Investigation 3: Telling Stories from Line Graphs*

Activity

Matching Numbers, Stories, and Graphs

The object of this activity is to match up some graphs, charts of plant heights, and stories about a plant's growth.

Briefly present an example to the class before they get started. Sketch a graph on the board, for example:

This line shows how a plant grew. How would you describe its growth? (At first it grew slowly; then it started to grow more quickly; then it slowed down and stopped growing.) **What might be some plant heights that could result in this graph?**

List their suggestions on the board. A reasonable sequence of numbers, for example, might be 2 cm, 3 cm, 4 cm, 7 cm, 10 cm, 11 cm, 11 cm.

When you think most of your students have the idea, pass out copies of Student Sheet 7, Matching Numbers, Stories, and Graphs, as well as scissors, some glue or tape, and an extra sheet of paper for mounting the work. Students cut out the graphs, sets of plant heights, and stories, and then match them up, making three complete sets. (Every set needs a chart of heights, a story, and a graph. There is an extra graph that doesn't belong with any of the stories or sets of heights.) They then glue or tape the sets together on a sheet of paper.

❖ **Tip for the Linguistically Diverse Classroom** As an aid to comprehension, act out each plant story on Student Sheet 7. While a student reads each story slowly, demonstrate what happens with real plants, drawings, or pictures. For example, to help explain the first story, you might draw a window with a shady tree outside, then a window with the sun directly outside.

When students finish, those that have time could make up another graph, set of plant heights, and story that go together.

Graphs, Stories, and Number Sequences ■ **69**

Students generally have the most difficulty choosing the chart of heights that goes with a graph and story. If this is a problem for some students, suggest that they use the numbers to find the amount of growth each day.

Where is the largest change in the number list? What is happening to the plant at that time? Where is the fastest growth shown on the graph?

Discussing the Matched Sets   When most students have finished, they get together in groups of four or five to go over the matching of each graph, story, and number sequence. If there is disagreement in a group, encourage students to explain their ideas. To resolve differences about matching number sequences and graphical shapes, it may be helpful for students to construct their own graphs from the sequence of heights in question.

Briefly bring the class together at the end of the session to resolve any disagreements that remain and to share what students learned in doing the activity and in discussing it.

Displaying Students' Plant Graphs   Students may conclude their plant measuring and graphing activities at this point. They might display chart and graph together on poster paper with a written statement of how their plant grew. If there is not time in class, they could finish this task as homework.

A.	Your plant was growing quickly for a while. Then you forgot to water it for several days. That made it grow more slowly.	Heights 2 4 6 8 9 10 11

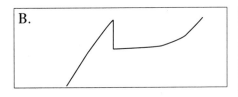

	Your plant was growing quickly for a few days. Then you dropped it and the top of it broke off. It stopped growing for a while before it started growing again.	Heights 1 3 5.5 8.5 6 6 7

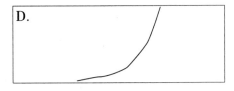

	Your plant was growing very slowly on a window sill that got no sunlight. You moved it to a sunny window. Then it started growing more quickly.	Heights 1 1.5 2 2.5 3.5 5 7

Session 5

Interpreting Graphs

What Happens

Students look at a variety of graphical shapes that represent the growth patterns of different plants and describe in words how each plant grew. Their work focuses on:

- interpreting different graphical shapes

Materials

- Student Sheet 8 (1 per student)

Activity

Distribute Student Sheet 8, Plant Graph Stories (a three-page handout), and introduce the task.

Each of the graphs on these three pages tells a story of how a different plant grew. You'll notice that there are no numbers on the graphs. What's important is their shapes. What does the shape of each graph tell you about how the plant grew at the beginning and how it grew later?

The questions after each graph ask you to write something about the plant's story. After you have interpreted all the graphs shown here, you will make a graph of your own.

Students work individually on this task. Circulate to make sure they understand the instructions for each section of the student sheet.

Assessment

Telling Stories from Graphs

❖ **Tip for the Linguistically Diverse Classroom** For questions 1–3 on Student Sheet 8, students who are not proficient in English can explain their answers through nonverbal responses, or you might ask a series of questions to break down the problem: "How does the line look when the plant is growing well? How does the line look when the plant slows down?" Sketch one of the graph sets on the board and ask students to add a third graph that shows a plant growing faster—or slower—than the two plants shown.

Students might answer questions 4 and 5 with a storyboard format. Then read question 6 aloud and work with students as they construct their graphs.

When everyone has finished, have a discussion with the whole class to share ideas. Encourage everyone who has a different idea to explain his or her thinking.

Use this assignment as well as the rest of the work in this investigation to assess students' understanding of graphs. For specific points to evaluate in each of the six problems on Student Sheet 8, see the following **Teacher Note,** Assessment: Telling Stories from Graphs.

Teacher Note — Assessment: Telling Stories from Graphs

Consider these points as you assess students' work on Student Sheet 8, Plant Graph Stories.

Problem 1 Do students associate the height of the line on the graph with the height of the plant? Correct answers include any that point out that plant B has grown more because its line goes higher.

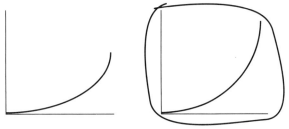

It is closer to the top than the other one.

Problem 2 Do students recognize that the farther across the graph a line goes, the more time has passed? Correct answers include any that point out that plant C grew faster because it grew to be the same height in a shorter time. Its line is steeper than plant D's line.

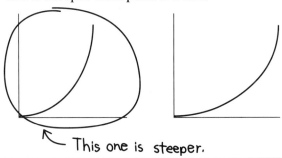

This one is steeper.

Problem 3 Do students associate steepness of the graph with rate of growth? Plant F is growing faster because the upward slope of the graph is greater. However, some students say that plant E is growing faster because it is taller. These students may be thinking about the time before that shown on the graph; they are considering the total growth of the plant, instead of the growth over the time interval indicated by the graph.

You may be able to clarify this by asking students to compare *your* growth during the last year to theirs. You grew a lot earlier in your life, but they are now growing fast and you are not growing at all. Their growth curve now would be an upward slant, while yours would be a flat horizontal line. Accept as correct the response of a student who chooses plant E but acknowledges that it grew more earlier and is not growing faster during the time shown on the graph.

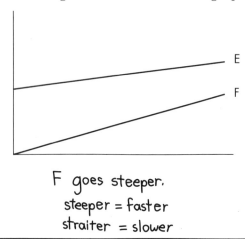

F goes steeper.
steeper = faster
straiter = slower

---

72 ■ *Investigation 3: Telling Stories from Line Graphs*

Problem 4 Can students distinguish between a graph that shows growth at a steady rate (the graph is a straight line) and one that shows a change in the rate of growth (the graph is curved)? Correct responses take notice of the ways growth rate changed for plants X and Z and didn't change for plant Y. For example, here are two sets of accurate student responses:

> Plant X got taller every day and then got taller faster than before.
>
> Plant Y got taller every day.
>
> Plant Z got taller fast and then got taller more slowly.
>
> Plant X grew slowly at first and then faster and faster.
>
> Plant Y grew pretty fast the whole time.
>
> Plant Z grew fast at first and then slowed down.

Problem 5 Do students understand that a line going up means that growth is increasing and a line going down means that growth is decreasing?

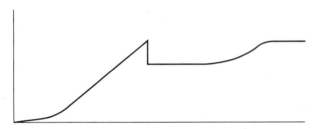

In this graph, students are asked to notice a number of changes in growth rate, including a sudden decrease in size. The best answers will mention the most changes in growth rate. For example, this fourth grader's explanation takes into account all the important changes in the plant's growth rate:

> The plant started slow then went fast. Then all of a sudden it dropped (it most likely fell off). It didn't grow for a few days. Then it started up slow. Then it went faster and then slowed down and stopped.

This next explanation fails to mention the significant decrease in growth:

> When the plant started to grow, it grew steady. Then it grew fast, then slow, then steady again.

The following explanation is even weaker, as it does not take as many important features into account:

> It went fast, then it shrunk, then it went fast and grew good.

Problem 6 Do students associate steepness of a graph with rate of growth and a horizontal line with no growth? A graph that fits the description for problem 6 would look something like this:

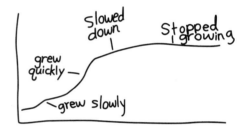

The best responses include those that accurately connect the steepness of the graph with the phrases telling how the plant is growing.

Sessions 6 and 7

Mystery Graphs

Materials

- Paper 11" by 17" or larger (2 pieces per group)
- Light and dark pens (1 each per group)
- Student Sheet 9 (1 per student, plus 1 cut up)
- Posted copy of list on Student Sheet 9

What Happens

Students make line graphs that show how the population changes throughout a day in different (unnamed) places or activities. The class tries to match the mystery graphs to a list of places and activities. Students revise their graphs to clarify anything that was confusing, and make a display to post outside the classroom. Their work focuses on:

- making and interpreting different graphical shapes

Activity

Making Graphs of Changing Population

Explain that students are going to make "mystery graphs." These graphs will show how the population of a certain place or a certain activity in your local area changes over 24 hours, on a school day. Then others in the class are going to try to guess which graph goes with each place or activity. These should be simple line graphs without numbers, similar to the plant graphs on Student Sheet 8.

Remind students about the shape of the home and school population graphs you discussed earlier. Draw these shapes on the board, listing some times along the horizontal axis.

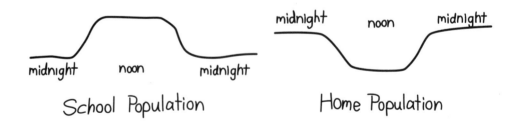

Point out that you are not showing exact numbers of people, and students should not use actual numbers in their mystery graphs, either. The graphs should just compare times when there are many people, few people, or no people.

74 ■ *Investigation 3: Telling Stories from Line Graphs*

Split the class into small groups of 3–4 students to make the mystery graphs. (Students need not work with the same partners from earlier sessions.) Hand out two large sheets of paper to each group. Suggest that they use a pencil to make a rough draft first. Then they can go over the lines with a light-colored marking pen so the graph can be seen from the front of the room. (They will use darker marking pens to make corrections later.)

Suggest that students place the paper with the longer dimension horizontal. They draw a line near the bottom of the paper for a horizontal axis. Along the line, they write 12:00 midnight at each end and 12:00 noon in the middle. They may write in other hours if they wish. They draw a vertical axis and write next to it only the words *number of people*. They do not write *any* numbers along it.

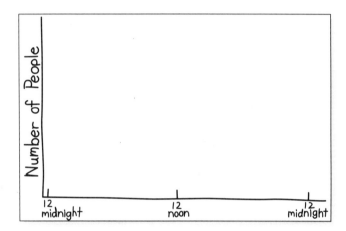

Hand each group of students a slip of folded paper with the name of a place or an activity from the cut-up copy of Student Sheet 9. Tell students not to let other groups know what they will be graphing, and not to put titles on their graphs. After they look at the piece of paper, they should put it away, out of sight.

Allow your students only 10 or 15 minutes to make their graphs. Warn them when there are only 2 or 3 minutes left to work.

While students are working, try to get around to every group. To encourage students to think about whether their graphs show precisely what they mean to show, you might make comments like the following. If what you describe is not true, they may need to rethink the shape of their line.

Here the line slopes up gradually between about 3 o'clock and 9 o'clock in the morning. More and more people must have started this activity all through that six-hour period of time.

Here the line goes up fast around 6 o'clock in the evening. It looks as if a lot of people start that activity all at once.

This hump is almost twice a high as this one. Were there twice as many people there at that point?

On this graph, the line never goes all the way to the bottom. There is always someone there.

There are two humps at about 12 noon and 9 o'clock. Those must be the times when a lot of people were there.

This line is pointed at the top. The most people were there for just a short while.

This graph has several high spots. People came and went in several groups.

When most groups are nearly finished, begin to collect the finished graphs and post them at the front of the room. For identification, write a different letter on each graph.

Activity

Discussing the Shapes of the Graphs

As the class views their displayed work, focus attention on one graph at a time. Ask students to notice the shape of the curve. They are not to guess yet what place or activity is depicted, just to make observations about the shape of the graph and what they can deduce from it. To give students the idea, begin with some comments like those you made to individual groups about the shape of the graph and what it means. Ask the authors of each graph if what the class sees is what they intended to show. For a sample discussion, see the **Dialogue Box,** The Shape of Mystery Graphs (p. 81).

Making Small Revisions Students take back their graphs to revise them as needed. Hand out dark-colored pens. Students do not make new graphs, but put in new parts of lines with the darker pen and draw over the old parts of lines they want to keep. Even the most accurate of the graphs may need to have a small change, such as making the slope of a line steeper to show that people came or went during a shorter period of time.

Remind students that their goal is to make the graph communicate so well that their classmates can guess which activity it describes; they are not trying to stump the others. Tell students that they will have a chance to make a clean copy of the graph later, so they should focus now on correcting the graph to make it as clear and accurate as possible.

Activity

Matching Graphs to Activities

Collect and post the groups' corrected graphs. Then post the large list of Mystery Places and Activities, and hand out Student Sheet 9. (To make the task a little easier, you might tell students to cross off a few of the options that no one graphed.)

Students write the letters of the group-made mystery graphs on the blanks where they think they belong.

Telling the Stories Invite one group of students at a time to come up and show their graph. Ask the other students:

What place or activity do you think this is? Explain why you think so.

Students may be able to give more than one answer that they can support with a reasonable story. After some discussion, the group who made the graph tells what place it represents and explains the curve through the day. They should not go hour by hour, but show the overall trends. For example:

During the morning, some people go jogging before breakfast and before going to work. Then they go to work, but a bunch go jogging at lunchtime ...

Mystery Graphs ■ **77**

Call up another group and repeat this process. After the class knows about a few of the graphs, it will be easier to figure out all of them, so your questions for discussion can change a little. For example:

When are the most people doing this activity? What other graph is like this one? How are they similar? How are they different? How can you tell this is not [the subject of the similar graph]?

Before finishing this activity, ask students about the process they went through in planning their graphs.

To plan these graphs in your group, what things did you talk about? What things did you have to agree on? What things were hard to agree on?

Activity

Revising the Mystery Graphs

With their mystery graphs in front of them, the small groups plan how they will improve them. Introduce this activity briefly by discussing the shapes of the lines:

Now that you know about how curves can show slow change or fast change or no change at all, plan how you will make your mystery graph clearer than it was before.

Think of a place that fills up with people within a short time. (for example, school, concert hall) **How would you use a line to show that?**

What places do people enter over a longer period of time? (for example, museums, public transportation) **How would you show that?**

What kind of place or activity do people go to and from all day, so that about the same number of people are there over a long period of time? (for example, shopping) **How would you show that?**

Students make new graphs that show more clearly the changing population of people doing the activity. While they are working, ask them to clarify the times at which people are coming and going and how quickly or slowly the number of people is changing.

Plan with students how to make a display of their mystery graphs as a quiz for others in the school. Post the improved graphs, with letters for identification, in a hall or other public area. Put with them a list of all the places and activities represented, adding a few extra items to the list. You might post an answer key, telling which graph represents which place or activity, in an inconspicuous place nearby or just inside your classroom door.

Answer Key to Students' Mystery Graphs (p. 79)

Mystery Graph A: eating meals; Mystery Graph B: riding the city subway (note that P.M. times are shown first); Mystery Graph C: at a movie theater.

Students can help you write an explanation to post with the display. For example:

> These graphs show the changing numbers of people doing activities or visiting certain places during a day. Look at the list of possible activities and places. Guess which graph goes with which place or activity. The correct answers can be found [tell where].

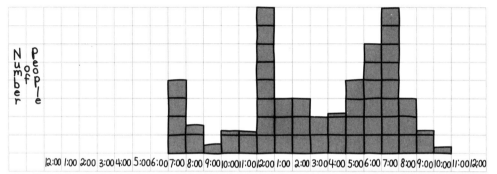

Mystery Graph A

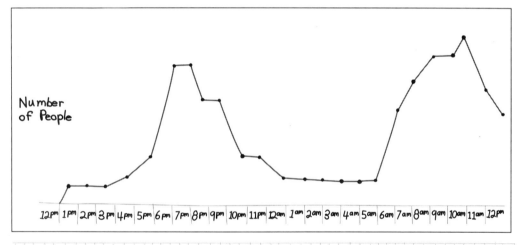

Mystery Graph B

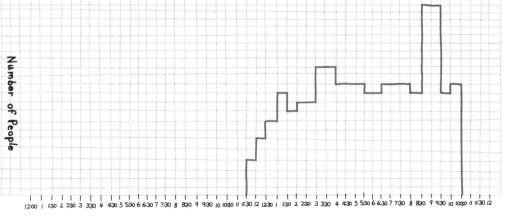

Mystery Graph C

Mystery Graphs ▪ **79**

Activity

Choosing Student Work to Save

As the unit ends, you may want to use one of the following options for creating a record of students' work on this unit.

- Students look back through their folders or notebooks and write about what they learned in this unit, what they remember most, what was hard or easy for them. You might have students do this work during their writing time.

- Students select one or two pieces of their work as their best work, and you also choose one or two pieces of their work, to be saved in a portfolio for the year. You might include students' written solutions to the assessments Revising Individual Graphs (p. 32) and Telling Stories from Graphs (p. 72). Students can create a separate page with brief comments describing each piece of work.

- You may want to send a selection of work home for parents to see. Students write a cover letter describing their work in this unit. This work should be returned if you are keeping a portfolio of mathematics work for each student.

Sessions 6 and 7 Follow-Up

Extensions

- **Similar Graphs for Different Activities** Pick one from among the students' mystery graphs and ask what other places that are *not* on the list might have graphs that look a lot like this graph. For example, if you picked the graph you did originally of the population at school, what other places would have such a flat line through the middle of the day?

- **Graphing Human Growth Curves** Make a transparency or copies of a graph of children's typical growth. Students can compare the story of human growth with the story of their plants' growth.

  Before seeing the published graph, students might sketch what they think the curve of their own growth would look like. Allow them a few minutes to sketch a graph of their own heights since they were born. Since they do not have access to an actual chart of their heights, they will need to think about when they probably grew the fastest, slowest, and most steadily.

  Volunteers can present their graphs and ideas to the rest of the class. When does their graph show the greatest change? The least? Then students can sketch what they think their future growth will look like.

  Finally, show the transparency or hand out copies of the graph of human growth. What do students notice? What surprises them? At what ages does a child grow the most? When is a child about half of his or her adult height?

DIALOGUE BOX

The Shape of Mystery Graphs

After groups have finished drawing the first draft of their mystery graphs, you hold a whole-class discussion of the general shapes, without touching on the subject of any graph. The point of this discussion is to help the groups clarify their graphs. Student observers are able to be quite specific about the shape of a graph, and this is sometimes very helpful to the authors.

Kenyana: I see highs in the afternoon and at 7 at night.

How would you describe the line after 7 o'clock?

Kenyana: It gradually goes down.

Nick: Gradually? It's pretty drastic.

Ahmad: It goes drastically down, then back up.

Rebecca: We made a mistake. The people come in at 7 and stay there. Can I fix my mistake? [*Rebecca takes her graph and draws the line after 7 o'clock at the same height as 7.*]

Look at [another graph]. What is different about this one?

Rafael: It goes up and down a lot. A lot more coming and leaving. Most late at night.

Where is the high in the morning?

Rafael: Between 7 and 8. The low is between 1 and 5.

When are the most people there?

Teresa: All the peaks are about the same, but they make sort of a clump at night. I guess that's more people than in the morning.

[*Speaking to the authors of the graph*] **So we see a small clump of people doing your activity between 7 and 8 in the morning. In the evening sometimes lots of people are there and sometimes very few. Is that what you wanted to show?**

Ten-Minute Math

Quick Images

Basic Activity

Students are briefly shown a picture of a geometric design or pattern, then draw it by developing and inspecting a mental image of it.

For each type of problem—2-D designs or dot patterns—students must find meaningful ways to see and develop a mental image of the figure. They might see it as a whole ("it looks like a four pointed star"), or decompose it into memorable parts ("it looks like four triangles, right side up, then upside down, then right side up, then upside down"), or use their knowledge of number relationships to remember a pattern ("there were 4 groups of 5 dots, so it's 20"). Their work focuses on:

- organizing and analyzing visual images
- developing concepts and language needed to reflect on and communicate about spatial relationships
- using geometric vocabulary to describe shapes and patterns
- using number relationships to describe patterns

Materials

- Overhead projector
- Overhead transparencies of the geometric figures you will use as images for the session; we have provided two transparency masters to get you started. To use the images on the masters, first make a transparency, then cut out the separate figures and keep them in an envelope. Include the numbers beside the figures because they will help you properly orient the figures on the overhead.
- Pencil and paper

Procedure

Step 1. Flash an image for 3 seconds. Show a picture of a geometric drawing or a dot pattern. (See below for specific suggestions related to these two options.)

It's important to keep the picture up for as close to 3 seconds as possible. If you show the picture too long, students will build from the picture rather than their image of it; if you show it too briefly, they will not have time to form a mental image. Suggest to students that they study the figure carefully while it is visible, then try to build or draw it from their mental image.

Step 2. Students draw what they saw. Give students a few minutes with their pencil and paper to try to draw a figure based on the mental image they have formed. After you see that most students' activity has stopped, go on to step 3.

Step 3. Flash the image again, for revision. After showing the image for another 3 seconds, students revise their drawing, based on this second view.

It is essential to provide enough time between the first and second flashes for most students to complete their attempts at drawing. While they may not have completed their figure, they should have done all they can until they see the picture on the screen again.

When student activity subsides again, show the picture a third time. This time leave it visible, so that all students can complete or revise their solutions.

Step 4. Students describe how they saw the drawing as they looked at it on successive "flashes."

Variations

In this unit you will find transparency masters for two types of Quick Images: geometric designs and dot patterns. You can supplement any of these with your own examples or make up other types.

Quick Image Geometric Designs. Use the Quick Image Geometric Designs transparency. When students talk about what they saw in successive flashes, many students will say things like "I saw four triangles in a row." You might suggest this strategy for students having diffi-

Ten-Minute Math

culty: "Each design is made from familiar geometric shapes. Find these shapes and try to figure out how they are put together."

As students describe their figures, you can introduce correct terms for them. As you use them naturally as part of the discussion, students will begin to use and recognize them.

Quick Image Dot Patterns Use the Quick Image Dot Patterns transparency. The procedure is the same, except that now students are asked two questions: "Can you draw the dot patterns you see? Can you figure out how many dots you saw?"

When students answer only one question, ask them the other again. You will see different students using different strategies. For instance, some will see a multiplication problem, 6 × 3, and will not draw the dots unless asked. Others will draw the dots, then figure out how many there are.

Using the Calculator You can integrate the calculator into the Quick Image Dot Patterns. As you draw larger or more complex dot patterns, students may begin to count the groups and the number of groups. They should use a variety of strategies to find the total number of dots, including mental calculation and the calculator.

Related Homework Options

- **Creating Quick Images** Students can make up their own Quick Images to challenge the rest of the class. Talk with students about keeping these reasonable—challenging, but not overwhelming. If they are too complex and difficult, other students will just become frustrated.

- **Family Quick Images** You can also send images home for students to try with their families. Instead of using the overhead projector, they can simply show a picture for a few seconds; cover it up while members of the family try to draw it; then show it again, and so forth. Other members of the family may also be interested in creating images for the student to try.

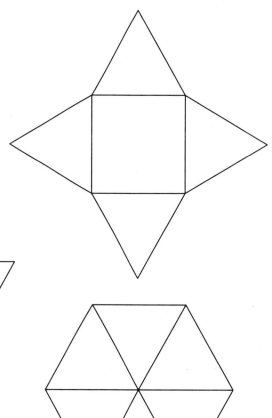

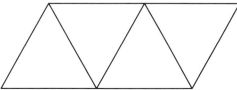

Quick Image Geometric Designs

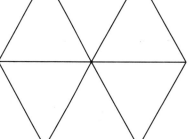

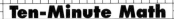

Broken Calculator

Basic Activity

Students work to get an answer on their calculator display while pretending that some of the keys are missing. The missing keys can be operations, numbers, or both. After students find one solution, they find others by making a small change in the first one. In this way, the solutions form a pattern.

Broken Calculator helps students develop flexibility in solving problems. They pull numbers apart and put them back together in a variety of ways as they look for expressions to substitute for given numbers. Students focus on:

- finding alternative paths to an answer when a familiar one isn't available
- finding many ways to get one answer
- writing related problems

Materials

Calculators: 1 per student

Procedure

Step 1. Pose the problem. For example, "I want to make 35 using my calculator, but the 3 key and the 5 key are broken. How can I use my calculator to do this task?"

Step 2. Students solve the problem by themselves. They record their solution in some way that another student can understand. Students in small groups check each others' solutions on their calculators.

Step 3. List some of the students' solutions on the board.

Here are some possible solutions to making 35 without the 3 and 5 key:

61 – 26   29 + 6   4 × 9 – 1

Step 4. Students choose one solution and extend it, making a series of related solutions. For example:

61 – 26	29 + 6	2 × 18 – 1
62 – 27	28 + 7	4 × 9 – 1
64 – 29	27 + 8	6 × 6 – 1
	26 + 9	
	24 + 11	

Students check each others' solutions and find another solution that follows the same pattern.

Variations

Restricting Number Keys

- Students make numbers without using the digits in those numbers, for example:

Make 1000 without using a 1 or a 0.

998 + 2
997 + 3
996 + 4

- Students make decimals without using the decimal point. Start with the simplest ones (0.1, 0.5, 0.25, 0.75, or 1.5) only after students have some experience relating them to fractions and division. You might start by providing a solution or two and challenge them to find some more: "I can make 0.5 on my calculator by using the keys 1 ÷ 2. Why do you think that works? Can you find another way to make 0.5?"

Some solutions for making 0.5 are as follows: 2 ÷ 4, 3 ÷ 6, 4 ÷ 8, 5 ÷ 10, 100 ÷ 200, 1000 ÷ 2000.

Restricting Operation Keys

- Students make a number using only addition. If you suggest a large number, students can make use of landmark numbers. For example:

Make 2754.

2000 + 700 + 54   2750 + 4   2749 + 5
2000 + 600 + 154   2751 + 3   2748 + 6

Ten-Minute Math

- Students make a number using only subtraction. The +, ×, ÷ keys are broken. Patterns of solutions for making 8 might look like these:

20 − 12	1008 − 1000
19 − 11	908 − 900
18 − 10	808 − 800
17 − 9	708 − 700

- Students make a number using only multiplication and division. The + and − keys are broken. Pick numbers that have many factors. Answers for making 24 might be:

1 × 24	24 ÷ 1	24 × 1 ÷ 1
2 × 12	48 ÷ 2	24 × 2 ÷ 2
3 × 8	72 ÷ 3	24 × 3 ÷ 3
4 × 6		24 × 4 ÷ 4

(One student filled a page with the third series so he could say he'd gotten the most answers.)

Restricting Both Operations and Digits

Make the missing operations problems more challenging by also not allowing students to use any of the digits in the final number. For example:

Make 654 using only addition and subtraction, and without using the digits 6, 5, or 4.

Related Homework Option

Pose one or two Broken Calculator problems only. Challenge students to solve the problems in more than one way, and to make their different solutions follow a pattern. They should write down their solutions so that another student can read them and know what to do on the calculator.

If students do not have calculators at home, give them time to try out their solutions the next day in school.

VOCABULARY SUPPORT FOR SECOND-LANGUAGE LEARNERS

The following activities will help ensure that this unit is comprehensible to students who are acquiring English as a second language. The suggested approach is based on *The Natural Approach: Language Acquisition in the Classroom* by Stephen D. Krashen and Tracy D. Terrell (Alemany Press, 1983). The intent is for second-language learners to acquire new vocabulary in an active, meaningful context.

Note that *acquiring* a word is different from *learning* a word. Depending on their level of proficiency, students may be able to comprehend a word upon hearing it during an investigation, without being able to say it. Other students may be able to use the word orally, but not read or write it. The goal is to help students naturally acquire targeted vocabulary at their present level of proficiency.

We suggest using these activities just before the related investigations. The activities can also be led by English-proficient students.

Investigation 1

in, out

1. Model going out the classroom door and then back in as you say:

   I am going out. I am coming in.

   Have a few students do the same.

2. Introduce simple drawings to represent in and out. For example:

   Have students point to them as you or other students go in and out.

3. Ask a few students to predict the number of times they can go in and out through the classroom door in 20 seconds. Have them test their predictions, counting ins and outs to see how close they come.

Investigations 1 and 2

change, size, age, color, amount

Note: Through this unit, students establish a special definition of the word *change*. In this introductory language activity, the word is presented in a more general sense.

1. Show students the following examples of change. Articulate the kind of change being shown for each example.

   a. Change in size: Compare one 8½ × 11 sheet of paper with a sheet of paper torn in half.

   b. Change in age: Show a picture of yourself as a child and a recent photo.

   c. Change in color: Show two pieces of dark construction paper, one of which has been faded by the sun.

   d. Change in amount: Show an unopened bag of raisins or pretzels, and a half-empty bag of the same item.

2. Ask students to show their understanding of different kinds of change nonverbally, for example:

   Clap once when I point to a change in color. Clap two times when I point to a change in size.

Investigation 3

runner, marathon, wheelchair racer, speed, slow, fast, quickly, pace, gradually, steadily

1. Put on running shoes and pin on a racer's number. Then pantomime warming up for a race. Explain that you will be a runner in a marathon. To give students an idea of the distance covered in a marathon (26.2 miles), pick a location approximately that distance from the school and tell students that you will be running that far. Supplement this introduction with photographs and videotapes of marathon races, if available.

2. Show a picture of a person in a wheelchair, if possible someone in or preparing for a race. Explain that people in wheelchairs take part in many athletic events, particularly in long-distance racing.

3. Explain that, since this is a very long race, you will go at different speeds during different parts of the race. Run in place as you demonstrate and explain different paces.

   I am now going slow. I am gradually speeding up. Now I am going faster.

   Then pantomime the same speeds as if you were in a wheelchair race, or have an English-proficient student pantomime simultaneously, one of you representing a runner and the other a wheelchair racer.

4. Challenge students' comprehension of these words by asking volunteers to act as if they were racers. Have them run in place, or pretend to wheel with their arms, at whatever pace you call out, changing pace as you tell them to.

   Run at a steady pace. Gradually slow down. Now speed up.

Blackline Masters

Family Letter
Student Sheet 1, Getting Data About Home Population
Student Sheet 2, Plant Height Chart
Student Sheet 3, Ways to Show Change
Student Sheet 4, The Boston Marathon
Student Sheet 5, Fast and Slow Growth
Student Sheet 6, Our Growing Plants
Student Sheet 7, Matching Numbers, Stories, and Graphs
Student Sheet 8, Plant Graph Stories
Student Sheet 9, Mystery Places and Activities
Motion Graph
Quick Image Dot Patterns
Quick Image Geometric Designs
One-Centimeter Graph Paper

_____ , 19 ____

Dear Family,

In mathematics class, we are starting a unit called *Changes Over Time.* We will be observing many different things that change over time and finding and inventing ways to illustrate such changes. Your child will have two activities that you might enjoy helping with at home.

1. In the first week, your child will need to figure out and write down the changing number of people who are at home over a 24-hour period, on a typical weekday. Your child is to write down when people come in and when they go out. At school, all the children will make graphs that show the changes in population in their homes during a day. They will look for patterns—when there are many people at home, and when there are just a few.

2. In the second week, your child will be looking for items from newspapers and magazines that show how things change. These could be charts, lists, pictures, cartoons, graphs, several days' worth of weather maps, or even headlines and other statements, such as "Increase in Kindergarten Enrollment Seen." At school the children will sort these items to see how different kinds of change are illustrated. The clippings will be shared and may be used on posters, so do not send anything to school that you need to have returned.

In school, your child will plant a bean seed and keep track of and graph its height over time. Looking at their graphs later, the children will try to answer questions like these: Does it grow at the same rate its whole life? How can you tell on the graph when it is growing fastest?

Learning about and describing changes like these is one of the important mathematical ideas that children will work on more and more as they go through school. We hope that you will participate in these activities, and that you will find them interesting.

Sincerely,

Getting Data About Home Population

At school, you are going to make a graph that shows the number of people who are at your home at different times over a 24-hour day. As homework, you need to collect the information for your graph.

Choose a typical weekday, when the people in your house do their usual activities.

Write down the times people usually go out and return home. Do not count quick trips to a nearby store or going outside to play.

You may count things such as going out to a friend's house, going out jogging, or delivering newspapers—but count them only if they happen for more than half an hour, at least two days each week.

Think about how you will make a graph that tells the story of the comings and goings at your home. The reader should be able to see quickly when most people are at home and when no one is at home.

Name _____   Date _____

Student Sheet 2

Plant Height Chart

Plant Name _____

Date	Day of the week	Height	Comments

Changes Over Time

Name _____ Date _____

Student Sheet 3

Ways to Show Change

Things are changing all the time. The weather changes. Prices go up and down. People get older. Cities get bigger. Animal populations change.

We have lots of ways to show such changes. See how many you can find. Look in newspapers. Look in magazines. Look for photographs, drawings, cartoons, ads, graphs, headlines, and anything else that shows how things are changing.

People at home can help you look. Talk about the things you find. Ask for permission to cut them out and bring them to school.

Name _____   Date _____

Student Sheet 4

The Boston Marathon (page 1 of 2)

Part 1 On the next page is the story about Bobby Hall's experience in the Boston Marathon, racing in a wheelchair. Use this copy of the story with the graph you made of the racer's speed.

The underlined words in the story describe the racer's speed and how it changed throughout the race. Find the places on your graph where the racer is doing what the underlined words describe. Write the words on your graph in the right place.

Part 2 Here is a list of the underlined words in the story.

    wheeled very fast   gradually slowing down
    slowed down a bit   kept pushing slowly
    wheeled nice and steady   picked up the pace
    push faster and faster   ran my fastest
    stopped   wheeled around slowly
    wheeled steadily

Some of these words and phrases describe the racer's speed. Some of them describe a *change* in speed. Sort these words and phrases into two lists. Put each word or phrase under the right heading below.

         Speed   Change in Speed

The Boston Marathon (page 2 of 2)

Participating in the Boston Marathon—all 26.2 miles of it—is an incredible experience. You're with thousands of other people, going through all kinds of different towns and cities and college campuses. Wheelchair participants get a lot of attention at Boston. People know us and they start us right at the front.

In the beginning of the race, there's about 4 miles of downhill. Like most wheelchair participants, I <u>wheeled very fast</u> on that downhill part. It felt good. But at mile 4, I remembered to pace myself. I <u>slowed down a bit.</u> For the next 9 miles or so, I <u>wheeled nice and steady.</u>

But then a very exciting thing happened. At Wellesley College, there was a huge crowd of students lining the course, screaming and clapping like crazy. All that excitement made me <u>push faster and faster</u> through the mile-long part of Wellesley. But then I noticed I was tired—too tired for being just a little more than halfway through the race. I <u>stopped</u> for a few seconds to get some water and pour some over my head. (I was getting hot, too!) I knew the hardest part of the race was coming up.

After Wellesley, I <u>wheeled steadily</u> for a few more miles. But then, by the seventeenth mile, it started getting hard. From mile 17 to 21 or so, I could feel myself <u>gradually slowing down.</u> There's a bunch of hard hills, and I knew I just had to take it easy to make it over those hills. My arms were aching so much. But the funny thing was, even once I had made it over the hills, I kept pushing slowly. I think by 21 miles I was running out of steam. It was hard to keep picking up my arms. Between 21 and 25 miles, I <u>kept pushing slowly</u>.

By 25 miles, I knew I would make it and I <u>picked up the pace.</u> The crowd was tremendous in the last mile or so. They just wouldn't let you slow down. The final stretch of a quarter mile or so is downhill, and I actually <u>ran my fastest</u> for that stretch. My arms felt all beaten up and shaky at the finish—but I <u>wheeled around slowly</u> for a while afterwards. That helps you keep from getting so stiff the next day.

— Bobby Hall (from a telephone interview)

Student Sheet 5

Fast and Slow Growth

Day	Height
Thursday	3 cm
Friday	4 cm
Monday	10 cm
Tuesday	13.5 cm
Wednesday	15 cm
Thursday	16 cm
Friday	16.5 cm
Monday	17.5 cm

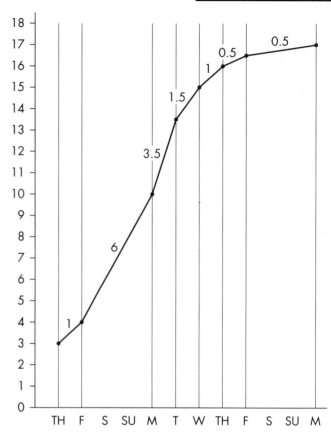

1. a. Where does this graph show the *fastest* growth?

   b. How can you tell from the *shape* of the graph?

   c. How can you tell from the *numbers* along the graph?

2. a. Where does the graph show *slowest* growth?

   b. How can you tell from the *shape* of the graph?

   c. How can you tell from the *numbers* along the graph?

96 ▪ *Changes Over Time*

Name _____ Date _____

Student Sheet 6

Our Growing Plants

Use this sheet with the graph you have made of your growing plant.

1. On your graph of your growing plant, write the changes (in centimeters) on each section of the line.

2. **a.** Find a place on your graph where your plant was growing fast. Write the word *fast* along your graph in this place.

   b. Explain how the *shape* of the graph tells you this is fast growth.

   _____

   _____

   c. Explain how the *numbers* on the graph tell you this is fast growth.

   _____

   _____

3. **a.** Find a place on the graph where your plant was growing more slowly. Write the word *slow* in this place.

   b. Explain how the *shape* of the graph tells you this is slow growth.

   _____

   _____

   c. Explain how the *numbers* on the graph tell you this is slow growth.

   _____

   _____

Changes Over Time ■ **97**

Student Sheet 7

Matching Numbers, Stories, and Graphs

Cut out the charts, stories, and graphs. Group them to make three sets that match.

Heights	Heights	Heights
1	1	2
3	1.5	4
5.5	2	6
8.5	2.5	8
6	3.5	9
6	5	10
7	7	11

Your plant was growing very slowly on a window sill that got no sunlight. You moved it to a sunny window. Then it started growing more quickly.

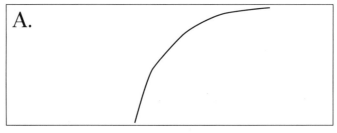

Your plant was growing quickly for a while. Then you forgot to water it for several days. That made it grow more slowly.

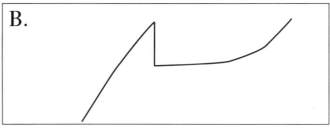

Your plant was growing quickly for a few days. Then you dropped it and the top of it broke off. It stopped growing for a while before it started growing again.

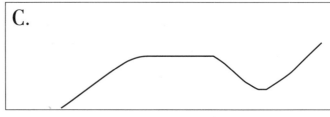

98 ■ *Changes Over Time*

Plant Graph Stories (page 1 of 3)

1. Plant A   Plant B

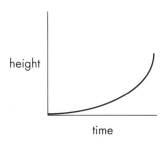 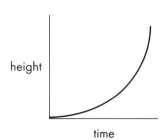

Which plant has grown more so far, A or B?
Explain how you decided.

_____

_____

2. Plant C   Plant D

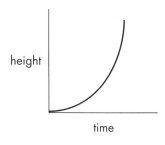 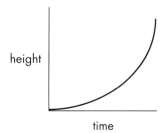

Which plant has grown faster so far, C or D?
Explain how you decided.

_____

_____

Plant Graph Stories (page 2 of 3)

3. **Plants E and F**

   Which plant is growing faster during the time shown on the graph, plant E or plant F? Tell how you decided.

   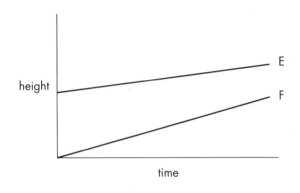

   _____
   _____
   _____

4. Tell the growth story of each plant below.

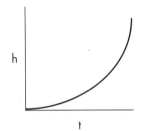

Plant X

_____
_____
_____

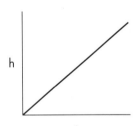

Plant Y

_____
_____
_____

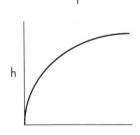

Plant Z

_____
_____
_____

Plant Graph Stories (page 3 of 3)

5. What might have happened to this plant? Tell the story of how it grew.

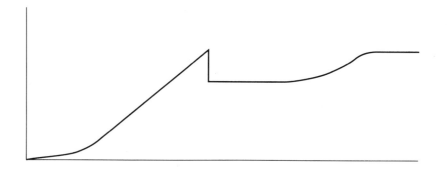

_____
_____
_____
_____
_____

6. Draw a graph that describes the following story:

   A plant <u>grew slowly</u> for 2 or 3 days and then <u>grew quickly</u> for 2 or 3 days. After this fast growth, it <u>slowed down</u>. In a few days, it <u>stopped growing</u>. It had reached its full height.

   After you have made your graph, write the underlined words along it in the right places.

Name _____ Date _____

Student Sheet 9

Mystery Places and Activities

_____ eating meals

_____ riding on public transportation

_____ at a movie theater

_____ watching TV

_____ jogging

_____ at a pizza restaurant

_____ at a public library

_____ talking on the telephone

_____ on the beach in warm weather

Motion Graph

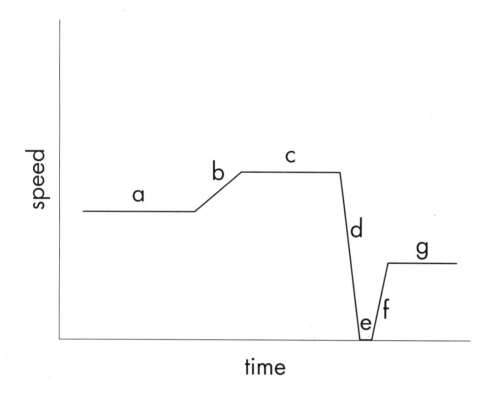

This graph shows a racer's speed during a race.

When is the racer speeding up?

When is the racer stopped?

When is the racer going at a slow, steady speed?

What is happening at d?

What is happening at f?

What is happening at c?

What might have happened to this racer during the race? Tell the whole story.

Quick Image Dot Patterns

1.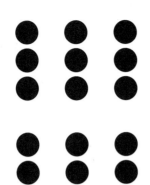

2.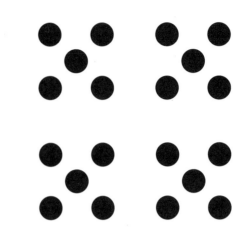

3.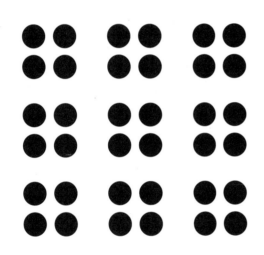

4.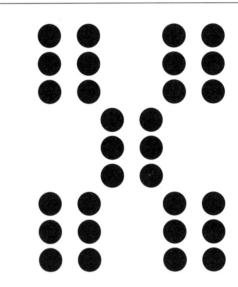

5.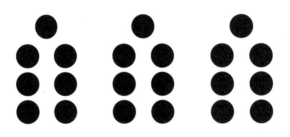

6.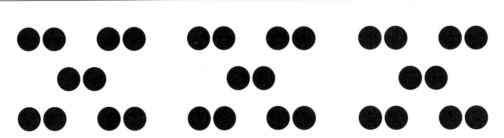

Quick Image Geometric Designs

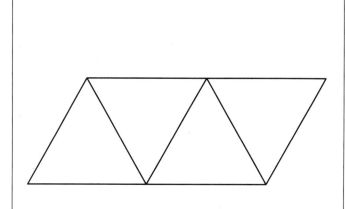

1.

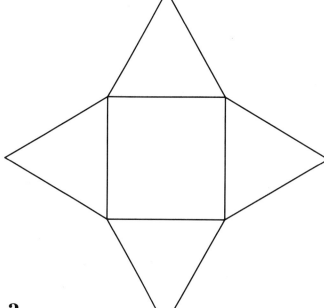

2.

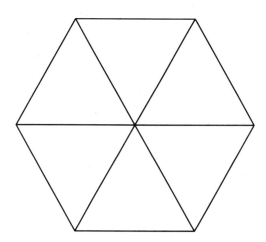

3.

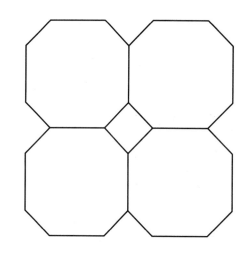

4.

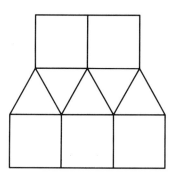

5.

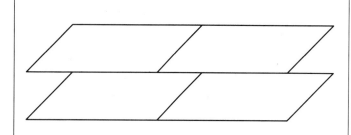

6.

ONE-CENTIMETER GRAPH PAPER

© Dale Seymour Publications